SpringerBriefs in Immunology

More information about this series at http://www.springer.com/series/10916

Parag Jain • Ravindra Pandey
Shiv Shankar Shukla

Inflammation: Natural Resources and Its Applications

Springer

Parag Jain
Department of Pharmacology
Columbia Institute of Pharmacy
Raipur, Chhattisgarh, India

Ravindra Pandey
Department of Pharmacognosy
Columbia Institute of Pharmacy
Raipur, Chhattisgarh, India

Shiv Shankar Shukla
Department of Analytical Chemistry
Columbia Institute of Pharmacy
Raipur, Chhattisgarh, India

ISSN 2194-2773   ISSN 2194-2781 (electronic)
ISBN 978-81-322-2162-3   ISBN 978-81-322-2163-0 (eBook)
DOI 10.1007/978-81-322-2163-0
Springer New Delhi Heidelberg New York Dordrecht London

Library of Congress Control Number: 2014957960

Printed on acid-free paper

Springer is part of Springer Science+Business Media (www.springer.com)

Preface

Humans have been using natural resources for various ailments from their existence, and it is nature that supports and heals by the different ways and fulfills the needs of human beings. The human beings of different era in different parts of the globe independently discovered the drugs from natural sources.

Interestingly, the last few decades have witnessed a quantum jump in relevant and useful publications, especially with regard to books pertaining to medicinal plants and their uses for scientific fraternity and commoners. The primary emphasis of this book was placed on basic pharmacological principles of inflammation. Details and individual herbal drug properties were deliberately omitted in the interest of making drug action more transparent and affording an overview of the pharmacological basis of drug therapy for inflammation.

The authors wish to reiterate that *Inflammation: Natural Resources and Its Application* cannot replace a textbook of inflammation or medicinal plants, neither can it replace a book on pain management, nor does it aim to do so. Rather, this little book is designed to arouse the curiosity of the pharmacological novice, to help students of medicine and pharmacy gain an overview of the inflammation and its management by natural sources, and to review certain bits of information in a concise format, and, finally to enable the experienced pain therapist to recall certain factual data.

The book essentially comprises six chapters wherein each chapter begins with collectively accepted information of the plant, animal and marine sources and its applications. We earnestly hope that the book will provide the kind of information that will interest those who work or plan to begin work in the area of inflammation and its management by natural resources.

Raipur, Chhattisgarh, India　　　　　　　　　　　　　　　　　　Parag Jain
15 October 14　　　　　　　　　　　　　　　　　　　　　　Ravindra Pandey
　　　　　　　　　　　　　　　　　　　　　　　　　　Shiv Shankar Shukla

Contents

Abbreviations

5-HIAA	5-Hydroxyindoleacetic acid
5-HT	5-Hydroxytryptamine
cAMP	Cyclic adenosine monophosphate
CD31	Cluster of differentiation-31
CGRP	Calcitonin gene related peptide
COX	Cyclooxygenase
CRP	C-reactive protein
CRP	C-reactive protein
DHA	Docosahexaenoic acid
EGCG	Epigallocatechin-3-gallate
EPA	Eicosapentaenoic acid
ESR	Erythrocyte sedimentation rate
ETs	Endothelins
HFCS	High fructose corn syrup
ICAM-1	Intercellular adhesion molecule-1
ICE	Interleukin-1β-converting enzyme
IFNs	Interferons
IL	Interleukin-1
LTA$_4$	Leukotrienes-A$_4$
LTs	Leukotrienes
MCP-1	Monocyte chemoattractant protein-1
NO	Nitric oxide
PAF	Platelet-activating factor
PG	Prostaglandins
PGD$_2$	Prostaglandin D$_2$
PGE$_2$	Prostaglandin E$_2$
PGF$_{2\alpha}$	Prostaglandin F$_{2\alpha}$
PGG$_2$	Prostaglandin G$_2$
PGH$_2$	Prostaglandin H$_2$
PGI$_2$	Prostaglandin I$_2$
PI	Phosphoinositide
PKC	Protein kinase C
PLA$_2$	Phospholipase A$_2$

PMN	Polymorphonuclear neutrophils
PV	Plasma viscosity
TNF-α	Tumor necrosis factor-α
Tx	Thromboxane
TXA$_2$	Thromboxane A$_2$
VCAM-2	Vascular cell adhesion molecule-1

List of Figures

List of Tables

About the Author

Parag Jain M. Pharm Pharmacology, Assistant Professor in the Department of Pharmacology, Columbia Institute of Pharmacy, Raipur. He completed his master's degree in 2013 from Chhattisgarh Swami Vivekanand Technical University (Columbia Institute of Pharmacy, Raipur). He has been honored by Indian Academy of Sciences (IASc), Bangalore, in 2012 for promotion towards research and selected for P.C. Dandiya prize for poster presentation on anti-inflammatory activity of Artesunate in 2013. He was involved in cutting-edge research work at Indian Institute of Integrative Medicine (IIIM), CSIR, Jammu, for 6 months and All India Institute of Medical Sciences (AIIMS), New Delhi, for 2 months during his master's, where he got expertise in toxicity testing, preclinical studies, pharmacokinetics, and pharmacodynamics, with equal inclination to ethnopharmacology. His areas of interest include neuropharmacology, toxicology, molecular biology, diabetes, ethnopharmacology and so on. He has recognized publications in national and international journals to his credit and few more in the pipeline. His present work lies on exploring the natural sources for treating type-2 diabetes.

Dr. Ravindra Pandey Ph.D. Pharmacognosy, Associate Professor in the Department of Pharmacognosy, Columbia Institute of Pharmacy, Raipur, India, is a renowned academician in the field of medicinal plants. His specific area of research is phytochemistry, natural products, and herbal medicines. He has more than 10 years of teaching and professional experience. He has several research papers published in national and international journals. He is an approved examiner and paper setter of different Indian Universities. He has a longstanding interest in teaching at the PG level and is involved in taking courses in pharmacognosy, microbiology, and isolation and standardization of herbal drugs.

Dr. Shiv Shankar Shukla Ph.D. Technology, is a Professor in the Department of Pharmacognosy, Columbia Institute of Pharmacy, Raipur. He completed his Ph.D. in the development of quality control parameters of some formulations containing herbals in 2012. His areas of research interest are spectroscopy and fingerprinting of herbal formulations. He has 7.5 years of teaching experience. He has been awarded with Young Scientist of Chhattisgarh award from Chhattisgarh Council of Science and Technology under the discipline of medical science and pharmaceutical science and Dr. P.D. Sethi annual award for best paper for his paper on TLC densitometric fingerprint of 6-Gingerol in 2010. He has more than 27 publications to his credit.

Introduction

Abstract
Nature is an inevitable source on earth; it may refer to the various types of living plants and animals and in some cases to the processes associated with inanimate objects. Natural resources include land, forests, wildlife resources, fisheries, water resources, energy resources, marine resources, and mineral resources. The way that these particular types of things exist and serve the humankind for better lifestyle is fairly noticeable. Forests provide us variety of services and nurture the mankind with its incomparable facilities and functions such as maintaining oxygen levels in the atmosphere, removal of carbon dioxide, control over water regimes, and slowing down erosion and also produce products such as food, fuel, timber, fodder, medicinal plants, etc. Marine fishes are an important protein food and contain significant medicinal value. There are several principles that each of us can adopt to bring about sustainable lifestyles. This primarily comes from caring for our Mother Nature and its resources in all respects. Regard for the nature is the greatest sentiment that helps bring about a feeling of protecting and conserving the natural resources, looking at how we use natural resources in a new and sensitive way. Through this book, we are going to explore the richness of nature, its kindness towards human race and purpose for the whole world.

1.1 Natural Resources

Natural resource is the form of matter or energy which is available in the earth and used by living things. Natural resources in the ecosystem are often characterized by amounts of biodiversity and geodiversity existent. These are essential for our survival and various valuable uses. These resources occurring naturally within environments may exist as air, water, plant, as well as living organism such as fish and microbes or may exist in other forms which may be processed to obtain the resources such oil, ores, and most forms of energy. Anything which comes from nature and

© Springer India 2015
P. Jain et al., *Inflammation: Natural Resources and Its Applications*,
SpringerBriefs in Immunology, DOI 10.1007/978-81-322-2163-0_1

people can use is natural resource; soil, sunlight, water, and wood are renewable resources, while coal, oil, and minerals are nonrenewable resources.

The living part of nature consists of plants and animals, including microbes. Plants and animals are biotic part of nature; but they require specific abiotic conditions to live in. The forests, grasslands, deserts, mountains, rivers, lakes, and the marine environment all form abiotic habitats for specialized communities of plants and animals and all are closely linked to each in their own habitat. Interactions between the abiotic aspects of nature and specific living organisms together form the basis of ecosystems.

1.1.1 Sources of Medicinal Value

Natural products have been playing an important role throughout the world in treating and preventing human diseases for a long time. Natural product medicines have come from various source materials including terrestrial plants, terrestrial microorganisms, marine organisms, and terrestrial vertebrates and invertebrates [1, 2]. Figure 1.1 represents the sources available against inflammation. Different natural sources used for anti-inflammatory and other medicinal purposes are stated below:

1.1.1.1 Plant Source
Plants have been catering as rich source of effective and safe medicines from ancient times. Plants have been used to attempt cures for diseases and to relive physical

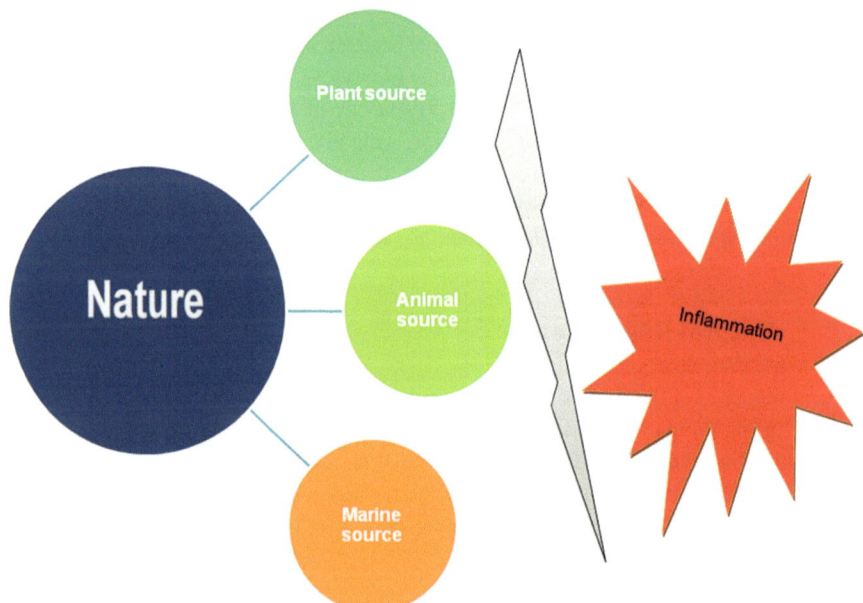

Fig. 1.1 Natural resources against inflammation

suffering. Plant's medicinal value is due to the presence of active chemical constituents or secondary metabolites, namely, alkaloids, glycosides, flavonoids, tannins, resins, terpenoids, saponins, etc. All these chemical substances are responsible for healing properties against several diseases. The large proportion of natural products in drug discovery has stemmed from the diverse structures and the intricate carbon skeletons of natural products. Since secondary metabolites from natural sources have been elaborated within living systems, they are often perceived as showing more drug likeness and biological friendliness than totally synthetic molecules [3]. Medicinal plants are source of raw materials for both traditional systems of medicine and modern medicine. These medicines are also in great demand in the developed world for primary health care because of their efficacy, safety, and lesser side effects. Indian Vedas describe the widespread use of herbal medicines and aqueous extracts of different plant parts for curing different diseases [4].

1.1.1.2 Animal Source

Animal source foods can provide a variety of micronutrients that are difficult to obtain in adequate quantities from plant source foods alone. Animals have been used as medicinal resources for the treatment and relieve of a myriad of illnesses and diseases in practically every human culture. The phenomenon of zootherapy represents a strong evidence of the medicinal use of animal resources. The healing of human ailments by using therapeutics that are obtained from animals or ultimately are derived from them is known as zootherapy [5].

Over 500 species of insects, mites, and spiders are used as medicines to cure both common and complicated ailments in Chhattisgarh, India [6]. Tetrameric polypeptide "melittin," the major component of bee venom, may be responsible for antiarthritic and anti-inflammatory effects [7]. Traditionally, leeches are usually used in conditions like abnormal swellings, piles, inflammatory abscess, skin diseases, rheumatoid arthritis, eye diseases, and poisonous bites.

1.1.1.2.1 Marine Source

Marine sources provide significant number of natural products with range of medicinal benefits. Marine natural products are obtained from major taxonomic groups like Mollusca, Arthropoda, Echinodermata, Porifera, Cnidaria, and many other phyla used in treatment of several diseases. Organisms from these source show wide range of therapeutic properties including anti-inflammatory, anticancer, antimicrobial, antihypertensive, anticoagulant, wound healing, and other medicinal effects [8]. The marine environment is a rich source of biologically active natural products of diverse structural types. The sponge *Luffariella variabilis* produces a chemical called monalide having anti-inflammatory activity by inhibiting the action of enzyme phospholipase A_2 [5].

Marine sponges belong to the genus Ircinia and are known to be a very rich source of terpenoids, several of which have shown a wide variety of biological activities. Variabilins, a polyprenyl hydroquinone, are chemicals obtained from sponges used as analgesic and anti-inflammatory properties [9].

Marine wealth exploration often leads to new ideas and new theories and discoveries, including new medicines. From slime molds to sponges, researchers are exploring the ocean's depths for new medications to treat inflammation, cancer, bacterial infections, viruses, heart disease, pain, and other ailments. Microbes having immense genetic and biochemical diversity look likely to become a rich source of novel effective medicines.

References

1. Newman DJ, Cragg GM, Snader KM (2000) The influence of natural products upon drug discovery. Nat Prod Rep 17:215–234
2. Paterson I, Anderson EA (2005) The renaissance of natural products as drug candidates. Science 310:451–453
3. Koehn FE, Carter GT (2005) The evolving role of natural products in drug discovery. Nat Rev Drug Discov 4:206–220
4. Meena AK, Bansal P, Kumar S (2009) Plants-herbal wealth as a potential source of ayurvedic drugs. AJTM 4(4):152–170
5. Costa-Neto EM (2005) Animal-based medicines: biological prospection and the sustainable use of zootherapeutic resources. An Acad Bras Cienc 77(1):33–43
6. Oudhia P (1995) Traditional knowledge about medicinal insects, mites and spiders in Chhattisgarh, India. Insect Environ 4:57–58
7. Bisset NG (1991) One man's poison, another man's medicine? J Ethnopharmacol 32:71–81
8. De Zoysa M (2012) Medicinal benefits of marine invertebrates: sources for discovering natural drug candidates. Adv Food Nutr Res 65:153–69. doi:10.1016/B978-0-12-416003-3.00009-3
9. Jirge Supriya S, Chaudhari Yogesh S (2010) Marine: the ultimate source of bioactives and drug metabolites. IJRAP 1(1):55–62

Inflammation

2

Abstract

Inflammation is a tissue-destroying process that involves the recruitment of blood-derived products, such as plasma proteins, fluid, and leukocytes into perturbed tissue. The source of the inflammation needs to be found by identifying the trigger factors which cause the inflammation, which includes physical, environmental, emotional, chemical, and most important nutritional factors. Inflammation cascade is regulated by immunological, physiological, and behavioral processes that are orchestrated by immune signaling molecule called cytokines. Inflammation is the basis for many diseases; it progresses to more serious degenerative conditions that ultimately get diagnosed as disease, i.e., irritable bowel syndrome, Crohn's disease, arthritis, allergies, hay fever, infertility, endometriosis, skin disorders like eczema along with major disorders including cancers, dementia, Alzheimer's disease, atherosclerosis, diabetes, and autoimmune disorders like systemic lupus erythematosus. While inflammation is necessary for debridement after injury, understanding the important role and stages of inflammation in the process of healing is fundamental to prevent chronic and debilitating diseases from occurring. This chapter signifies inflammation and its cause, mechanism, functions, and recovery.

2.1 Inflammation

Inflammation is derived from Latin word "inflamao" that means "Ignite, set alight" [1]. It is the process to remove the injurious stimuli like irritants, damaged cells, infection, and other stimulants and to initiate the healing process [2]. It may occur due to several causes such as burns, infections, chemical irritants, frostbite, toxins, physical injury, hypersensitivity, ionizing radiation, stress, trauma, and alcohol [2]. Several authors define inflammation in their own way but the basic phenomenon of inflammation is protecting body against injury. Authors outline inflammation as "Inflammation is a complex process that occurs in vascular tissues in response to

© Springer India 2015

P. Jain et al., *Inflammation: Natural Resources and Its Applications*,

SpringerBriefs in Immunology, DOI 10.1007/978-81-322-2163-0_2

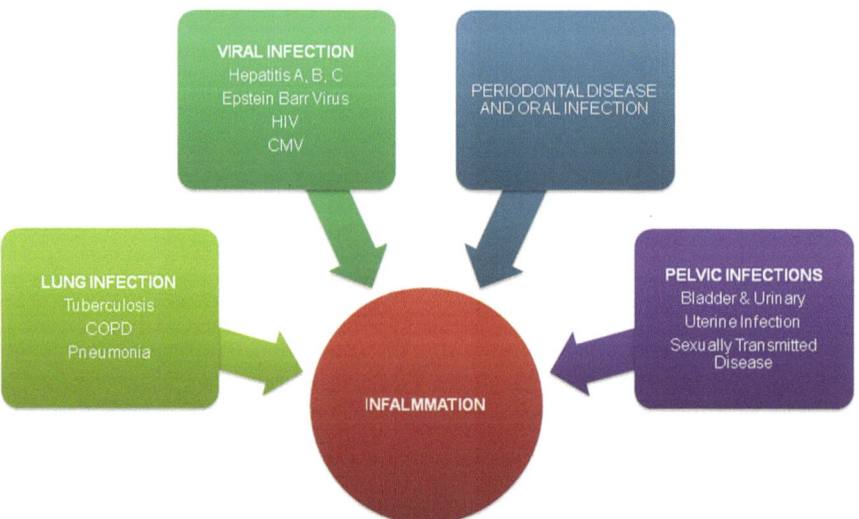

Fig. 2.1 Causes of inflammation

injurious stimuli like pathogens, damaged cells, or irritants." Or "Inflammation is a biological response that fills an area with blood in order to remove damaged tissue for recycling." Or "Inflammation is a first response of defense and healing that becomes active before our immune system does." Or "Inflammatory response is a series of events, or stages, that the body performs to attain homeostasis." Or "Inflammation is the body's physiological response or a nonspecific defense mechanism to a superficial cut, burn or a bacterial infection." Inflammation does not mean infection, even when an infection causes inflammation (Fig. 2.1).

2.2   Diagnosis of Inflammation

Inflammation in any body part makes release of extra protein from the site of inflammation which circulates in the bloodstream. The erythrocyte sedimentation rate (ESR), C-reactive protein (CRP), and plasma viscosity (PV) blood tests are commonly used to detect this increase in protein and so are markers of inflammation. Only correct diagnosis will relief from inflammation, while avoidance or incorrect diagnosis may give rise to more serious conditions like gangrene (Fig. 2.2).

2.3   Mechanism of Action

When a tissue is injured by any physical or chemical means, the process of inflammation starts automatically. Inflammation is caused by release of chemicals, most importantly prostaglandins (PGs), leukotrienes (LTs), histamine, bradykinin,

Fig. 2.2 Inflammation and diagnosis

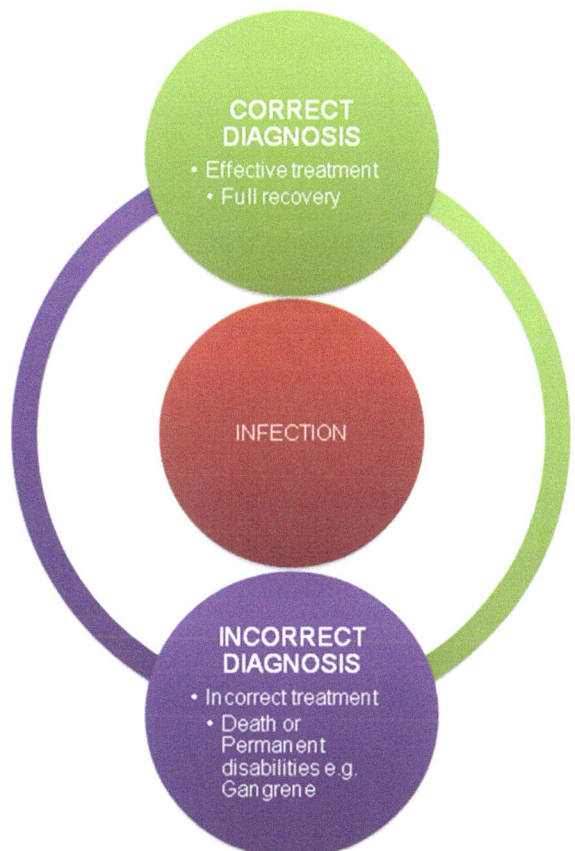

and, more recently, platelet-activating factor (PAF) and interleukin-1. All the antiinflammatory sources achieve their effects by blocking the increase in prostaglandins (PG) synthesis, competitively inhibiting histamine receptors. Aspirin-like anti-inflammatory drugs inhibit the cyclo-oxygenase (COX) enzyme and reduce synthesis of prostanoids. However, corticosteroids prevent the formation of both PGs and LTs by causing the release of lipocortin, which by inhibition of phospholipase A2 reduces arachidonic acid release.

2.4   Pathway of Inflammation

The inflammatory response is accompanied by erythema, edema, hyperalgesia, and pain. Inflammation is mediated by synthesis and release of various cytokines, growth transforming and chemotactic factors, and infiltration of various inflammatory cells mainly the neutrophils which try to protect from harmful effect of noxious stimuli and cause inflammation. Figure 2.3 represents the pathway of inflammation.

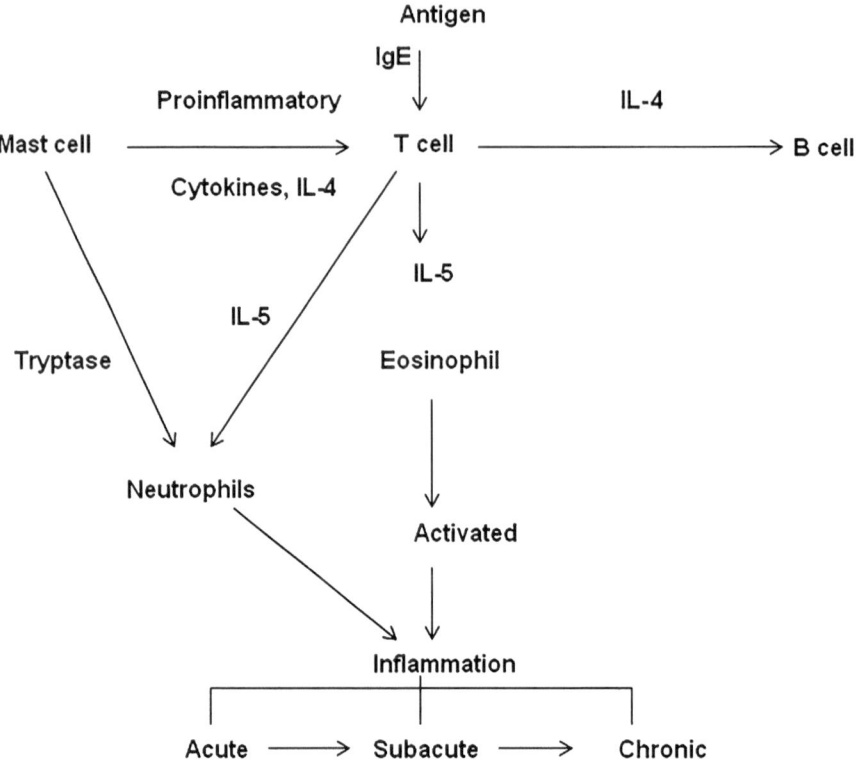

Fig. 2.3 Pathway of inflammation

2.5   Types of Inflammation

Broadly, inflammation is of two types, i.e., acute and chronic. Acute inflammation persists for a short duration from few hours to days, and the process is initiated by cells like macrophages, dendritic cells, histiocytes, Kupffer cells, and mastocytes. Acute inflammation comprises of two events, namely, the vascular and the cellular events. In chronic inflammation, the inflammatory process persists for a longer period and is generally accompanied by tissue granulation and fibrosis.

2.5.1   Acute Inflammation

Acute inflammation is the initial response of the immune system against pathogens and tissue injury which is characterized by ischemia, metabolic disturbance, and cell-membrane damage. It is a rapid self-limiting process, mediated by eicosanoids and vasoactive amines followed by vascular and cellular events which increase the movement of plasma and leukocytes into infected site. The proliferative phase of inflammation leads to granulation tissue formation lasts for 6–8 weeks. The final phase of acute inflammation deals with healing and scar formation. During this stage, cellular activity goes decrease and increased organization of extracellular matrix. Several inflammatory conditions are the result of acute inflammation, i.e., acute respiratory distress syndrome, acute transient rejection, asthma, glomerulonephritis, reperfusion injury, septic shock, and vasculitis.

2.5.1.1 Phases of Inflammation

Inflammation is immensely a complex response which seems to bring fluid, protein, and cells from the blood into the damaged tissues. It is known that the tissues dipped in watery fluid or extracellular lymph lack most of the proteins and cells that are present in blood, since most of the proteins are too large to cross the blood vessel endothelium. Thus, when damage and infection occurred, there is some mechanism which allows cells and proteins to expel out to extravascular sites. The inflammatory response includes vasodilatation, increased vascular permeability, cellular infiltration, and activation of cells of the immune system as well as of complex enzymatic systems of blood plasma (Fig. 2.4).

2.5.1.1.1 Vascular Events

The response which occurs in seconds of the tissue injury and last for minutes is acute vascular response. Vascular events comprise of vasodilatation and increased vasopermeability which leads to increased blood flow and entry of fluid into the tissues, and its duration varies in different types of inflammation [2]. This phase of inflammation response can be demonstrated by scratching the skin with a fingernail. The wheal and flare reaction which occurs is composed of various response blanching of skin due to vasoconstriction, appearance of red line due to capillary dilation, and flush due to arterioles dilation; a wheal appears as fluid leaks from the capillaries.

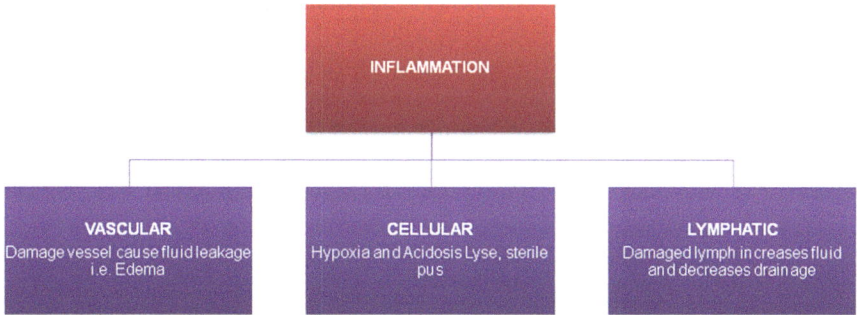

Fig. 2.4 Effects of inflammation on body system

2.5.1.1.2 Cellular Events

The acute cellular response takes place over the next few hours when sufficient damage to the tissue has occurred. The neutrophils appear in the tissues during this phase, erythrocytes also gain access into the tissues, and bleeding occurs. Platelets aggregate at the site of injury in which red cell stacked together help to stop bleeding and clot formation. The dead and dying cells contribute to pus formation followed by chronic cellular response which is characterized by appearance of mononuclear cell infiltrate composed of macrophages and lymphocytes. Macrophages are involved in microbial killing, in clearing up cellular and tissue debris. The cellular events begin due to movement of the leukocytes in the bloodstream and adhering to the endothelium by a process called margination. All these occur commonly at the site of postcapillary venules and take place because of the interaction of various adhesion molecules like selectins, integrins, ICAM-1 (intercellular adhesion molecule-1), VCAM-2 (vascular cell adhesion molecule-1), CD31 (cluster of differentiation-31), and Sialyl-Lewis X-modified proteins. The expression of these adhesion molecules is controlled by various cytokines [3]. After exiting the circulation, leukocytes migrate towards the site of injury by a process called chemotaxis in response to chemo attractants. These chemo attractants may be bacterial products, peptides, lipids, leukotrienes, and cytokines.

Over the next few weeks, resolution may occur, i.e., normal tissue architecture is restored. Blood clots are removed by fibrinolysis to return tissue to its original form. However, if it is not possible to remove the infectious agents which have accumulated at the site completely, they are walled off from the surrounding tissue in granulomatous tissue. A granuloma is formed when macrophages and lymphocytes accumulate around material that has not been eliminated, together with epithelioid cells and giant cells that appear later, to form a ball of cell.

2.5.2 Chronic Inflammation

Chronic inflammation is inflammation of prolonged duration that lasts for several months to years in which continuing inflammation, tissue injury, and healing, often by fibrosis, proceed simultaneously. Chronic inflammation may arise due to persistent infections by microbes like *Mycobacterium tuberculosis*, *Treponema pallidum* and certain viruses and fungi. These microorganisms elicit T lymphocyte–mediated immune response called delayed-type hypersensitivity. Autoimmunity also plays an important role in several common and debilitating chronic inflammatory diseases such as rheumatoid arthritis, inflammatory bowel disease and psoriasis [4]. Chronic inflammation also relates to some diseases like Alzheimer disease, atherosclerosis, metabolic syndrome, type 2 diabetes and some forms of cancer. Prolonged exposure to silica leads to silicosis, a chronic inflammatory condition. Figure 2.5 shows different types of inflammation according to its progression.

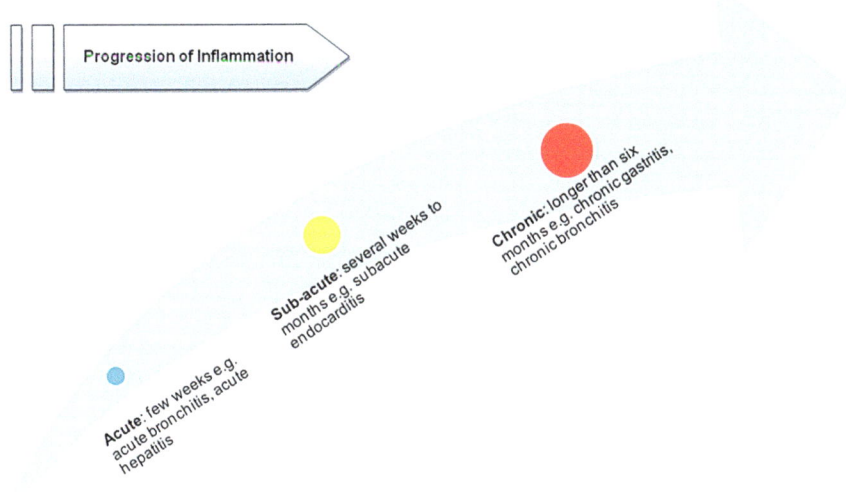

Fig. 2.5 Types of inflammation

2.6   Inflammatory Cells

Inflammatory cells participate in acute and chronic inflammation are polymorphonuclear neutrophils (PMN), basophils, eosinophils, lymphocytes, plasma cells, and monocytes that response to a foreign substance.

2.6.1   Mast Cells

Mast cells are an important source of a variety of proinflammatory mediators and cytokines than can promote inflammation and vascular changes. Main substances released from mast cells are histamine, heparin, leukotrienes, platelet-activating factor (PAF), nerve growth factor, and some interleukins. Therefore, they are considered to be involved in tissue repair [5, 6].

2.6.2   Polymorphonuclear Leukocytes

Neutrophil polymorphs are the shock troops of inflammation and are the first of the blood leukocytes to enter an inflamed area. Polymorphonuclear neutrophils (PMN) contain many substances in their granules like protease, myeloperoxidase, lysozyme acid, and alkaline phosphatase. PMN are considered to be first line of defence against bacteria because of phagocytosis of microorganisms and engulfment of antigen-antibody complex. The process is regulated by the adhesion molecules, i.e., selectins, intercellular adhesion molecule (ICAM), and integrins. The neutrophil is attracted to the invading pathogen by chemicals termed chemotaxins which are released by microorganisms or by local cells such as macrophages (e.g., chemokines). Neutrophils can engulf, kill, and digest microorganisms by generating toxic oxygen products and enzymatic digestion. Neutrophils are live, apoptotic, and constitute pus. When neutrophils have released their toxic chemicals, they undergo apoptosis and must be cleared by macrophages [5].

Eosinophils kill multicellular parasites like helminthes. These include eosinophil cationic protein, a peroxidase, the eosinophil major basic protein, and a neurotoxin. Eosinophils are considered in pathogenesis of asthma where granule proteins cause damage to bronchiolar epithelium. The number of eosinophils increases in conditions like allergy, parasitic infections, skin diseases, and certain lymphomas.

Basophils are found in certain parasitic infections and hypersensitivity reactions. They form only 0.5 % of circulating white blood cells.

2.6.3 Monocytes/Macrophages

Monocytes arrive in inflammatory lesions several hours after the polymorphs. Adhesion pattern of this cell is similar to neutrophils, although monocyte chemotaxis utilizes additional chemokines, such as MCP-1 (monocyte chemoattractant protein-1). Activated macrophages release many substances like collagenase, elastase, plasminogen activating factor, products of complement, coagulation factor, interleukin-1, tumor necrosis factor (TNF), and oxygen-derived free radicals that act on vascular endothelial cells, attract other leukocytes to the area, and give rise to systemic manifestations of the inflammatory response such as fever. Macrophages engulf tissue debris and dead cells, killing most microorganisms [6].

2.6.4 Vascular Endothelial Cells

Vascular endothelial cells play active role in inflammation; endothelial cells secrete nitric oxide, causing relaxation of the underlying smooth muscle, vasodilatation, and increased delivery of plasma and blood cells of the inflamed area. The endothelial cells regulate plasma exudation and delivery of plasma-derived mediators and express variety of receptors including those for histamine, acetylcholine, and interleukin-1 [6].

2.6.5   Platelets

Platelets are primarily involved in coagulation and thrombotic phenomena but also play a part in inflammation. They can generate free radicals and proinflammatory cationic proteins in addition to thromboxane A_2 and platelet-activating factor (PAF). They have low affinity receptors for IgE and are believed to contribute to the first phase of asthma.

2.6.6   Neurons

Some sensory neurons release inflammatory neuropeptides and are fine afferents with specific receptors at their peripheral terminals. Kinins, 5-hydroxytryptamine, and other chemical mediators generated during inflammation act on these receptors, stimulating the release of neuropeptides such as neurokinin A, substance P, and calcitonin gene-related peptide (CGRP) [6].

2.7   Inflammation and Recovery

Recovery of an inflammatory condition is a complicated process; it starts as soon as the injury occurs and can take hours, days, weeks, and months or more to complete. The inflammatory phase normally takes between 24 and 72 h, depending on the severity of damage. Injured tissue can include skin, muscles, ligaments, or tendons. The recovery or healing includes inflammatory phase, proliferation phase, and remodeling phase. Inflammatory phase is a great deal of activity in the tissues. Where there is bleeding, a blood clot forms to join tissue back together, at the same time vascular and cellular activity starts dealing with microbes, foreign bodies, and dying tissue. The inflammation process is instigated and controlled by chemical mediators released from blood, cells, and tissues. This includes histamine, Hageman factor, and bradykinin. The proliferation phase is next and can take up to 21 days. While the macrophage cells are busy cleaning up, they also instigate the proliferation stage by secreting active mediators, which include chemotactic and growth factors; these mediators propagate granulation tissue. Granulation tissue consists of fibroblasts which entangle with fibronectin molecules, contracts, and can reduce the wound area by up to 30 %. Lastly, the remodeling stage is also known as the maturation stage. This is the stage of wound strengthening in which realignment or remodeling of the collagen occurs, so that the bundles of fibers become aligned along the stress of the tissue. When the process is finally complete, the capillaries in the area are reduced and the collagen bundles increase in size. Figure 2.6 represents overall inflammation and recovery process.

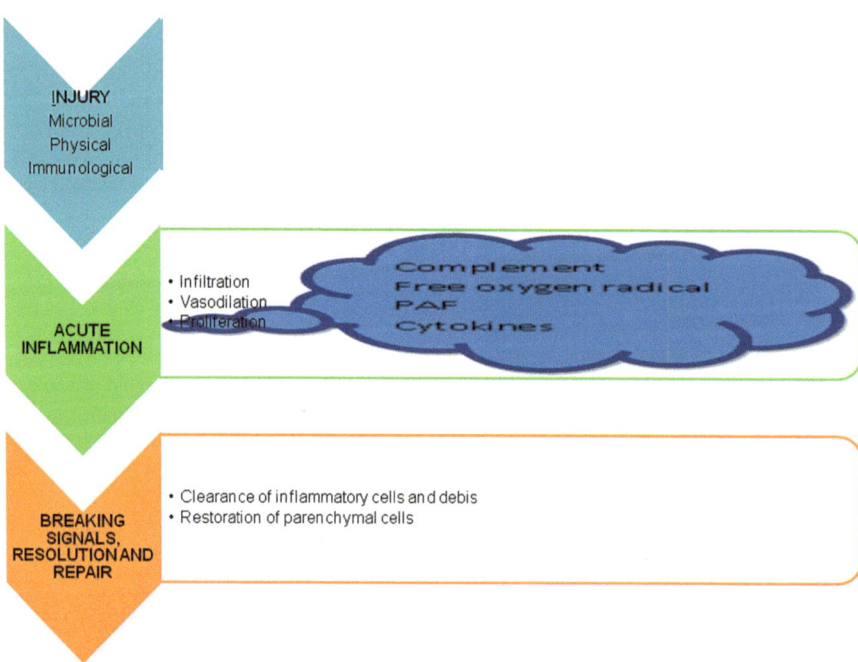

Fig. 2.6 Inflammation and recovery

References

1. Ferrero-Miliani L, Nielsen OH, Andersen PS, Girardin SE (2007) Chronic inflammation: importance of NOD2 and NALP3 in interleukin-1beta generation. Clin Exp Immunol 147(2):227–235
2. Ryan GB, Majno G (1977) Acute inflammation. A review. Am J Pathol 86(1):183–276
3. Lentsch AB, Ward PA (2000) Regulation of inflammatory vascular damage. J Pathol 190:343–348
4. Mohan H (2013) Textbook of pathology, 6th edn. Jaypee Brothers Medical Publishers, New Delhi
5. Eming SA, Krieg T, Davidson JM (2007) Inflammation in wound repair: molecular and cellular mechanisms. J Investig Dermatol 127:499–500. doi:10.1038/sj.jid.5700730
6. Rang HP, Dale MM (2007) Local hormones, inflammation and immune reactions. In: Rang HP, Dale MM, Ritter JM, Flower RJ (eds) Rang and Dale's pharmacology, 6th edn. Churchill Livingstone, London

Inflammatory Mediators

3

Abstract

Inflammation is the complex pathophysiologic response of tissue injury or infection. Biochemical mediators released during inflammation intensify and propagate the inflammatory response leading to organ dysfunction and major problem in many clinical conditions such as sepsis, severe burns, acute pancreatitis, hemorrhagic shock, and trauma. Inflammatory mediators are soluble, diffusible molecules that act systemically and locally at the site of injury or infection. Inflammation causes stimulation of body's defense system; activation of leukocytes causes release of inflammatory mediators at a site of infection or inflammation which control the later accumulation and activation of other cells. These mediators secrete primarily from blood plasma, neutrophils, monocytes, macrophages, platelets, mast cells, endothelial cells lining the blood vessels, and damaged tissue cells. This chapter summarizes recent studies that demonstrate the critical role played by inflammatory mediators in the pathogenesis of acute and chronic inflammation. It is reasonable to speculate that elucidation of the key mediators in inflammation, coupled with the discovery of specific inhibitors, would make it possible to develop clinically effective anti-inflammatory therapy.

3.1 Inflammatory Mediators

There are several inflammatory mediators, either cell derived or plasma derived, released at the time of tissue injury. Cell- and plasma-derived mediators work in concert to activate cells by binding specific receptors, activating cells, recruiting cells to sites of injury, and stimulating release of additional soluble mediators. These mediators are relatively short-lived or are inhibited by intrinsic mechanisms, effectively turning off the response and allowing the process to resolve. The cell-derived mediators like histamine, serotonin, prostaglandins, leukotrienes, and platelet-activating factor are released at the site of inflammation, whereas the plasma-derived

© Springer India 2015
P. Jain et al., *Inflammation: Natural Resources and Its Applications*,
SpringerBriefs in Immunology, DOI 10.1007/978-81-322-2163-0_3

Table 3.1 Inflammatory mediators

Cell-derived mediators	Functions
Histamine	Arteriole dilation and increased venous permeability
Serotonin	Vasoconstriction in denervated tissue, vasodilatation in intact tissue
Prostaglandins	Vasodilatation, pain perception
Leukotriene	Leukocyte adhesion, potent chemoattractant, release of lysosomal enzymes
Cytokines	Vasodilatation, increased vasopermeability, and phagocytosis
Interleukins	Activation and chemoattraction of neutrophils
Plasma-derived mediators	Functions
Bradykinin	Vasodilatation, increase vascular permeability, smooth muscle contraction, and induce pain
Complement system	Stimulates histamine release by mast cells

mediators are mostly generated in the liver and circulate in the plasma. Histamine is one of the best-known mediators released from cells during inflammation which triggers vasodilatation and increases vascular permeability. The prostaglandins are a group of fatty acids produced by many types of cells, responsible for sensitizing pain at nerve ending. The sources of the principal mediators and their roles in the inflammatory reaction are summarized in Table 3.1.

3.1.1 Cell-Derived Mediators

The major cells involved in inflammatory responses include mast cells, neutrophils, eosinophils, macrophages, lymphocytes, endothelial cells, and platelets. These cells react to a variety of signals, stimuli, and irritants to orchestrate the inflammatory response. Cell-derived mediators have redundant activities and include variety of chemical groups such as vasoactive amines (histamine and serotonin), lipid mediators (prostaglandins and leukotrienes), and cytokines (interleukins and tissue necrosis factor). Mast cells, platelets, and basophils produce the vasoactive amines serotonin and histamine. Other mediators are derived from injured tissue cells or leukocytes recruited to the site of inflammation.

3.1.1.1 Histamine
It is a biogenic amine mediator formed by decarboxylation of the amino acid histidine by the enzyme L-histidine decarboxylase which is stored in granules of mast cells and basophils (Fig. 3.1). In mast cells and basophils, histamine is complexed in intracellular granules with an acidic protein and a high molecular weight heparin termed macro heparin. Histamine release is initiated by a rise in cytosolic Ca^{2+}. Agents that increase cAMP (cyclic adenosine monophosphate) formation, e.g.,

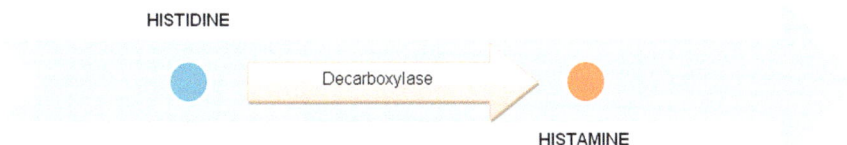

Fig. 3.1 Synthesis of histamine

β-adrenoceptor agonists, inhibit histamine secretion [1]. Histamine causes vasodilatation and increases permeability and endothelial gap formation. The release of cytokines, namely, interleukin-1 (IL-1) and tumor necrosis factor-α (TNF-α) by the macrophages and other cell types, triggers the release of the biogenic amines [2]. Histamine is a well-known mediator involved in anaphylaxis and causes severe edema, pruritus, and bronchospasm.

Histamine receptors are of two types H_1 and H_2, both are G protein-coupled receptor. H_3 receptor has been described recently which acts as inhibitory receptor on central nervous system. Histamine stimulates protein kinase C (PKC) via phosphoinositide (PI) hydrolysis through H_1 receptor activation. Increased cAMP level in lung fragments indicates that H_2 receptors are positively coupled to adenylyl cyclase in lung. In human skin, histamine causes a vasodilating response (flare) that is mediated by H1 receptors. Human bronchial vessels are relaxed by low concentrations of histamine in vitro but are constricted by high concentrations [3]. It is likely that the vasodilating response is the result of the release of nitric oxide (NO) from endothelial cells and that the vasoconstricting effect is the result of the direct action of histamine on vascular smooth muscle H1 receptors.

3.1.1.2 Serotonin

Serotonin [5-hydroxytryptamine (5-HT)] causes vasoconstriction in most animal species. It is formed by decarboxylation of tryptophan and is stored in secretory granules (Fig. 3.2). Degradation occurs mainly by monoamine oxidase, forming 5-hydroxyindoleacetic acid (5-HIAA), which is excreted in urine and serves as an indicator of 5-HT production in the body. Serotonin is derived from mast cells, but in humans, it is found in serum after blood clot [4]. It was subsequently found in the gastrointestinal tract and central nervous system (CNS) and shown to function both as a neurotransmitter and as a local hormone in the peripheral vascular system. Over 90 % of the total amount in the body is present in the enterochromaffin cells in the gut. In the blood, it is present in high concentration in platelets, and in CNS, it is present in localized region of the midbrain. These are divided into seven classes (5-HT$_{1-7}$), one of which (5-HT$_3$) is a ligand-gated ion channel and the rests are G protein-coupled receptors [1]. It causes increase in gastrointestinal motility, contraction of smooth muscle, platelet aggregation, stimulation of peripheral nociceptive endings, and excitation/inhibition of central nervous system.

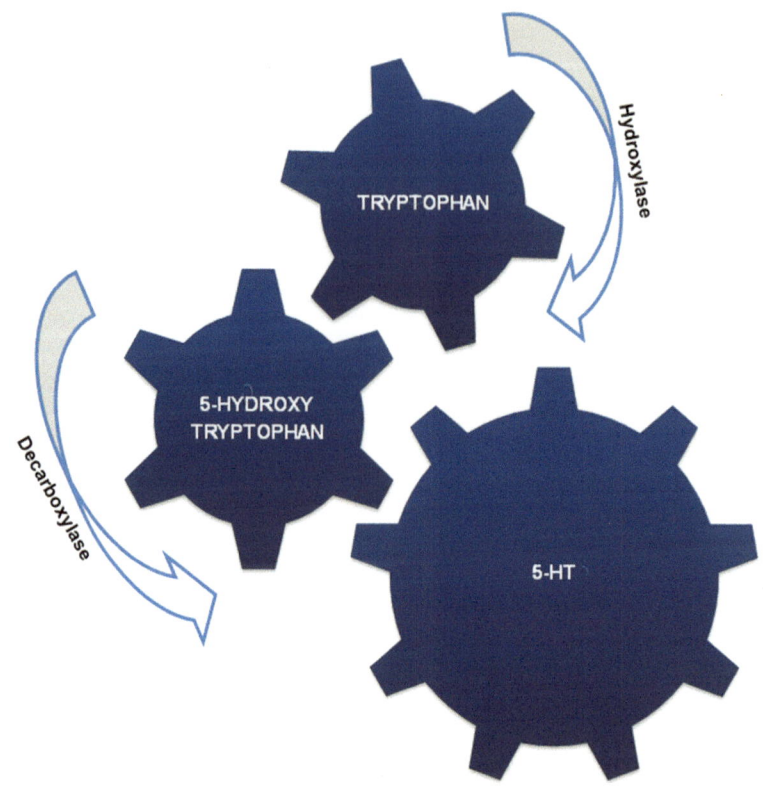

Fig. 3.2 Synthesis of 5-hydroxytryptamine

3.1.1.3 Adenosine

Adenosine is a purine nucleoside that is produced by dephosphorylation of 5′-AMP by the membrane-associated enzyme 5′-nucleotidase and is liberated intracellularly by cleavage of the high-energy bonds of adenosine triphosphate, adenosine diphosphate, and cyclic 5′-AMP. However, during hypoxia or even excessive cell stimulation, when the utilization of energy and oxygen exceeds the supply, 5′-AMP is metabolized to adenosine. It exists free in the cytosol of all cells and is transported in and out of the cells mainly using a membrane transporter. Most of the cells are capable of producing adenosine in times of energy deficit. Adenosine can be released by lung tissue in times of hypoxia, such as after allergen-induced bronchoconstriction, when the circulating levels of adenosine have been shown to be 3 times the baseline concentrations. Adenosine includes three receptor subtypes A_1, A_{2a} and A_{2b}, and A_3. These all are G protein-coupled receptor. Interaction of adenosine with these receptors leads to either inhibition of adenylyl cyclase (A_1), stimulation of adenylyl cyclase (A_{2a} and A_{2b}), or activation of phospholipase C (A_3) [1]. Functionally, A_1 causes blockade of AV conduction and reduction of force of conduction, neuroprotection against cerebral ischemia through inhibition of glutamate release, and bronchoconstriction. A_2 receptors deal with vasodilatation, inhibition of platelet aggregation, and stimulation of nociceptive afferent neurons especially in the heart. A_3 causes release of mediators from mast cells. A_3 receptors have been recently identified on human eosinophils, and activation of these receptors by adenosine inhibits eosinophil migration. Activation of A_3 receptors on eosinophils has also been shown to lead to an increase in $[Ca^{2+}]$.

3.1.1.4 Prostanoids

Prostanoids are generated from arachidonic acid by COX (Cyclooxygenase). It is of two types COX-1 and COX-2 [5]. Prostanoids include prostaglandins (PGs) and thromboxane (Tx). Arachidonic acid plays a central role in inflammation related to injury and many disease states. COX-1 is present in most cells as a constitutive enzyme that produces prostanoids that act as homeostatic regulators, whereas COX-2 is inducible by inflammatory stimuli, such as endotoxin and proinflammatory cytokines, and its induction is inhibited by glucocorticoids (Fig. 3.3a). Both enzymes catalyze the incorporation of two molecules of oxygen into every arachidonate molecule, forming the highly unstable endoperoxides prostaglandin G_2 (PGG_2) and prostaglandin H_2 (PGH_2).

These are rapidly transformed by isomerase or synthase enzymes to prostaglandin E_2 (PGE_2), prostaglandin I_2 (PGI_2), prostaglandin D_2 (PGD_2), prostaglandin $F_{2\alpha}$ ($PGF_{2\alpha}$), and thromboxane A_2 (TXA_2), which are bioactive end products of this reaction (Fig. 3.3b). There are five main classes of prostanoid receptors, all of which are typical G protein-coupled receptors. They are termed DP, FP, IP, EP, and TP receptors, respectively, depending on whether their ligands are PGD_2, $PGF_{2\alpha}$, PGI_2, PGE_2, or TXA_2 [6].

PGE_2 causes contraction of bronchial and gastrointestinal smooth muscles through action on EP_1, dilatory action on EP_2, contraction of intestinal smooth muscles, and inhibition of gastric acid secretion through action on EP_3 receptors. PGD_2 causes vasodilatation, inhibition of platelet aggregation, relaxation of gastrointestinal and uterine muscle, and release of pituitary hormones. $PGF_{2\alpha}$ causes myometrial contraction in humans, luteolysis, and bronchoconstriction in some species. PGI_2 causes vasodilatation, inhibition of platelet aggregation, renin release, and

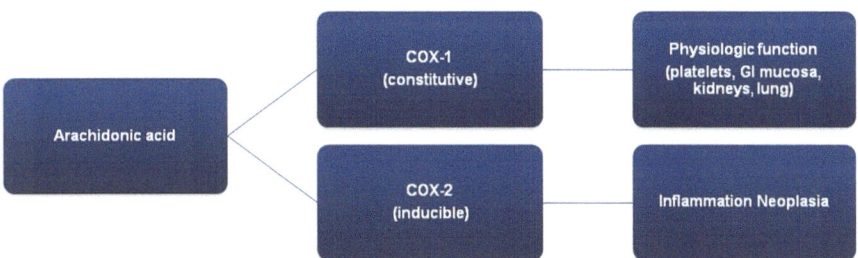

Fig. 3.3a Role of arachidonic acid

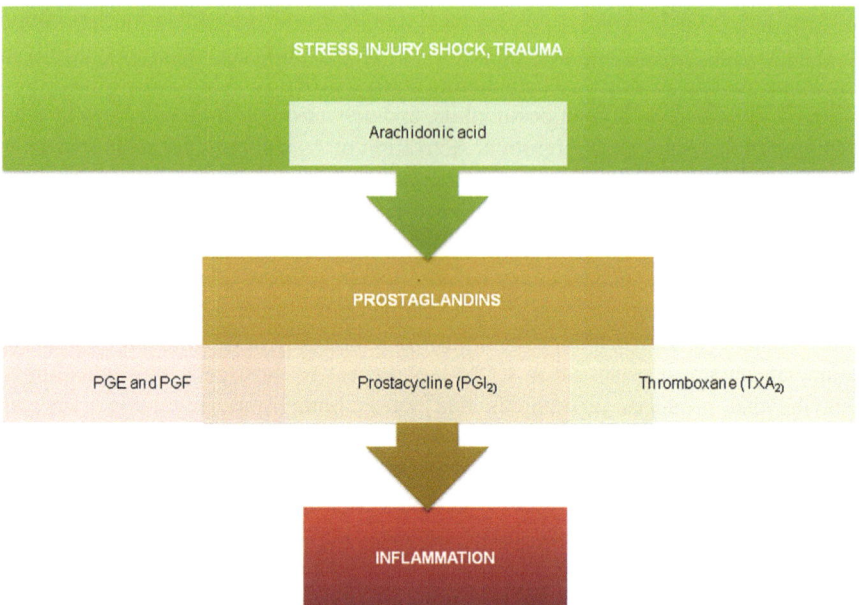

Fig. 3.3b Stimulation of inflammation

natriuresis through effects on tubular reabsorption of Na^+. TXA_2 causes vasoconstriction, platelet aggregation, and bronchoconstriction. In areas of acute inflammation, PGF_2 and PGI_2 are generated by the local tissues and blood vessels, while mast
cells release mainly PGD_2. In chronic inflammation, cells of the macrophage series
also release PGE_2 and TXA_2. Prostanoids have effects on the release of inflammatory mediators from inflammatory cells which inhibit the release of mediators from
mast cells, monocytes, neutrophils, and eosinophils. The receptors involved are
probably EP_2 receptors. PGE_2 favors the development of helper T (Th2) cells by
inhibiting interleukin-2 (IL-2) and interferon (IFN)-y production in human CD41
cells and inhibiting the secretion of interleukin-12 (IL-12) from macrophages [3].
Moreover, in presence of PGE_2, the dendritic cells cause Th2 cell differentiation and
increased synthesis of interleukin-5 (IL-5). It is suggested that the anti-inflammatory
action of prostanoids against eosinophils may predominate over its T cell action.

3.1.1.5 Leukotrienes

Leukotrienes (LTs) are synthesized from arachidonic acid by lipoxygenase-
catalyzed pathways (Fig. 3.4). The main enzyme in this group is 5-lipoxygenase. On
cell activation, this enzyme translocates to the nuclear membrane, where it associ-
ates with a crucial accessory protein affectionately termed FLAP (5-lipoxygenase-
activating protein) [3]. Several types of cells, including mast cells, eosinophils,

Fig. 3.4 Synthesis of leukotrienes

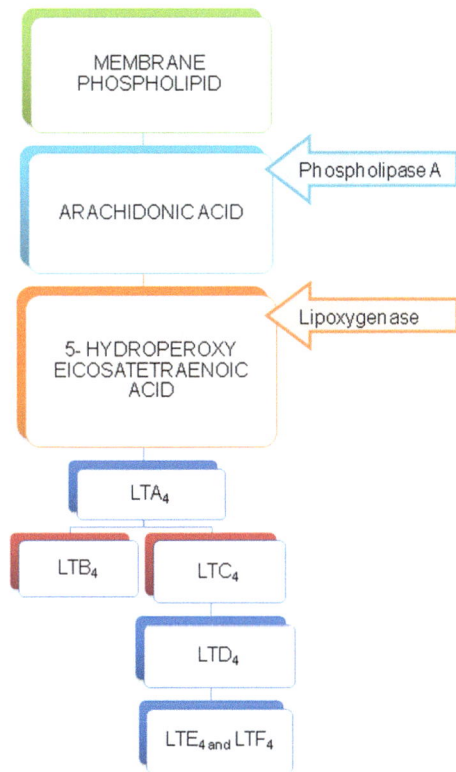

macrophages, neutrophils, and epithelial cells, can synthesize LTs in response to a variety of stimuli. The 5-lipoxygenase incorporates a hydroperoxy group at C5 in arachidonic acid, leading to the production of the unstable compound leukotrienes-A4 (LTA$_4$). This may be converted enzymically to LTB$_4$ and is also the precursor of the cysteinyl-containing leukotrienes LTC$_4$, LTD$_4$, LTE$_4$, and LTF$_4$ [1].

3.1.1.6 Platelet-Activating Factor

Platelet-activating factor (PAF) is a biologically active lipid that can produce effects at exceedingly low concentrations. It is believed to be an important mediator in both acute and chronic allergic and inflammatory phenomena. PAF is biosynthesized from acyl-PAF in a two-step process. The action of phospholipase A$_2$ (PLA$_2$) on acyl-PAF produces lyso-PAF, which is then acylated to give PAF. Large amounts of PAF can be synthesized by several inflammatory cell types in the lung, including resident cells such as mast cells and alveolar macrophages. The major enzyme responsible for the catabolism of PAF is PAF acetylhydrolase. It is now known to be an intracellular acetylhydrolase enzyme present in the cytoplasm of several inflammatory cell types, including mast cells, macrophages, and platelets. PAF is a G protein-coupled receptor and its cell signaling pathways include increases in [Ca^{2+}], increases in IP3 and diacylglycerol levels, and induction of cell cycle-active genes, such as *egr*-1, *fos*, and *jun* [3]. PAF produces vasodilatation, increased vascular permeability, and weal formation. PAF is a potent chemotaxin for neutrophils and monocytes and recruits eosinophils into the bronchial mucosa in the late phase of asthma. It can activate PLA_2 and initiates eicosanoid synthesis.

3.1.1.7 Endothelins

Endothelins (ETs) are potent constrictor peptides that were originally described as vasoconstrictors released from endothelial cells. There are three ET peptides known: ET_1, ET_2, and ET_3 [6]. ETs responses are mediated by at least two receptor subtypes ET_A and ET_B, both are G protein-coupled receptors. ETs may increase the release of inflammatory mediators from a variety of cells. ET-1 increases the release of lipid mediators from cultured human nasal mucosa and increases superoxide formation and tumor necrosis factor-α (TNF-α) release in alveolar macrophages [3]. ET-1 potently stimulates collagen secretion from pulmonary fibroblasts and therefore involved in the increased collagen formation observed in asthmatic airways.

3.1.2   Plasma-Derived Mediators

Mediators derived from plasma include complement and complement-derived peptides, kinins, and platelets. Released via the classic or alternative pathways of the complement cascade, complement-derived peptides (C3, C5a, and C5b) increase vascular permeability, cause smooth muscle contraction, activate leukocytes, and induce mast-cell degranulation. C5a is a potent chemotactic factor for neutrophils and mononuclear phagocytes. The kinins are also important inflammatory mediators. The most important kinin is bradykinin, which increases vascular permeability and vasodilatation and, importantly, activates phospholipase A2 to liberate arachidonic acid. Bradykinin is also a major mediator involved in the pain response. Platelets play only a reactionary role at the time of endothelial disruption and are now recognized as important mediators of the inflammatory process [6]. Platelets contribute as hemostasis and thrombosis during inflammation.

3.1.2.1 Bradykinin

Bradykinins are active peptides formed by proteolytic cleavage of circulating proteins termed kininogens through a protease cascade pathway. There are two bradykinin receptors, designated B_1 and B_2, both are G protein-coupled receptors and mediate very similar effects [1]. B_1 receptors are normally expressed at very low levels but are strongly induced in inflamed or damaged tissues by cytokines such as IL-1. It is likely that B_1 receptors play a significant role in inflammation and hyperalgesia, and its antagonists are now developing for neurological disorders. Bradykinin causes vasodilatation and increased vascular permeability. It is a potent pain-producing agent, and its action is potentiated by the prostaglandins. Bradykinin

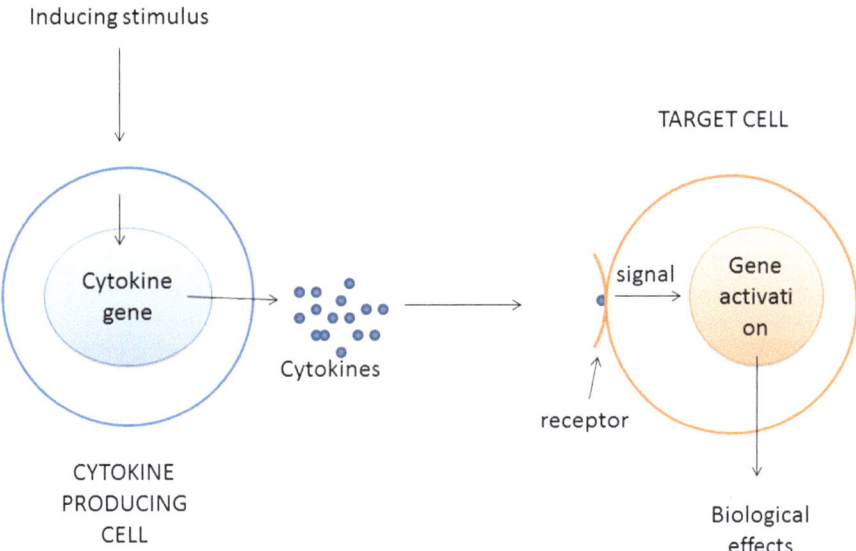

Fig. 3.5 Functioning of cytokines

also has spasmogenic actions on intestinal, uterine, and bronchial smooth muscle. Physiologically, the release of bradykinin by tissue kallikrein may regulate blood flow to certain exocrine glands and influence secretions.

3.1.2.2 Cytokines

Cytokine is an all-purpose functional term that is applied to protein or polypeptide mediators synthesized and released by cells of the immune system during inflammation. The family comprises interleukins, chemokines, interferons, colony-stimulating factors, growth factors, and tumor necrosis factors (TNFs). Chemokines act on G protein-coupled receptors, while the rests all act on kinase-linked receptors, regulating phosphorylation cascades that affect gene expression, such as the Jak/Stat pathway. Chemokines are defined as chemoattractant cytokines that control the migration of leukocytes, functioning as traffic coordinations during immune and inflammatory reactions. Cytokines have been linked to a complex signaling language, with the final response of a particular cell involved being determined by the strength and number of different messages received concurrently at the cell surface (Fig. 3.5). TNF-α and interleukin-1 (IL-1) are primary inflammatory cytokines which participate in acute and chronic inflammatory reactions as well as repair and resolution. Transforming growth factor-β (TGF-β), IL-4, IL-10, and IL-13 are anti-inflammatory cytokines that inhibit chemokine production [1].

References

1. Rang HP, Dale MM (2007) Local hormones, inflammation and immune reactions. In: Rang HP, Dale MM, Ritter JM, Flower RJ (eds) Rang and Dale's pharmacology, 6th edn. Churchill Livingstone, London
2. Rocha MF, Aguiar JE, Sidrim JJ, Costa RB, Feitosa RF, Ribeiro RA, Lima AA (2003) Role of mast cells and pro-inflammatory mediators on the intestinal secretion induced by cholera toxin. Toxicon 42(2):183–189
3. Barnes PJ, Chung KF, Page CP (1998) Inflammatory mediators of asthma: an update. Pharmacol Rev 50(4):520–544
4. Sirek A, Sirek OV (1970) Serotonin: a review. Can Med Assoc J 102(8):846–849
5. Soloman LM, Juhlin L, Kirschenbaum MB (1968) Prostaglandin on cutaneous vasculature. J Invest Dermatol 51:280–282
6. Steinhubl SR (2007) Platelets as mediators of inflammation. Hematol Oncol Clin North Am 21(1):115–21

Natural Sources of Anti-inflammation

4

Abstract

All the anti-inflammatory sources or drugs achieve their effects by blocking the increase in prostaglandin (PG) synthesis. PG synthesis is blocked by blocking the enzyme cyclooxygenase – an enzyme in arachidonic acid cascade for the synthesis of prostanoids. Mother Nature created these natural supplements eons ago, and it has been present in ecosystems for over a billion years. Hundreds of plant metabolites are reported to have many pharmacological activities although most of these reports are of academic interest and very few find entry at clinical trials. Compilation of the information would help promote wider acceptance and use of these nature-based drugs in mainstream of medicine. The present chapter is directed towards compilation of the pharmacological attributes of natural sources in the drug discovery and development process as it could be a driving force to identify lead molecules providing an attractive strategy for novel and improved therapeutics. In this chapter, we are going to reveal the name of these natural medicines and different sources available to us and what health benefits it provides to us to fight against inflammation.

4.1 Anti-inflammatory Source

Nature is the inevitable source on earth to fulfill our basic requirements for living. Plant drugs have been the major source for treatment of diseases for a long time. Alkaloid is a group of biological amine and cyclic compounds having nitrogen in the ring, naturally occurring in plant, microbes, animals, and marine organisms. Many plants, including the Amazon bark cats' claw (*Uncaria tomentosa*) and the common spice rosemary (*Rosmarinus officinalis*), demonstrate powerful anti-inflammatory, pain-relieving properties. In toxicity studies, the anti-inflammatory plants demonstrate great safety. Many hundreds of plants contain well-known anti-inflammatory agents. Hops (*Humulus lupulus*), a herb used in beer brewing, contains a group of compounds called the humulones, which are being studied for their

© Springer India 2015
P. Jain et al., *Inflammation: Natural Resources and Its Applications*,
SpringerBriefs in Immunology, DOI 10.1007/978-81-322-2163-0_4

significant anti-inflammatory and pain-relieving properties. The science of plant-based medicines is significantly advanced, and we now have products available to us that will reduce or eliminate inflammation without being hazardous to health. Herbs can be used as the sole therapy in autoimmune disease or as complementary corticosteroid sparing therapies allowing patients to take smaller doses or shorter courses of corticosteroids.

Alternately, marine medicines also play a role in inflammation apart from that plant-derived bioactive compounds showed anti-inflammatory actions. The marine environment associated with chemical diversity constitutes an actually unlimited resource of new active substances in the field of the development of bioactive products. Due to phenomenal biodiversity, the marine world is a rich natural resource of many biologically active compounds such as polyunsaturated fatty acids (PUFAs), sterols, proteins, polysaccharides, antioxidants, and pigments. Marine organisms include algae, sponges, coelenterates, bryozoans, mollusks, tunicates, echinoderms, miscellaneous marine organisms and marine microorganisms, and phytoplankton. Many marine organisms produce a wide variety of biologically active metabolites which cannot be found in other organisms. Moreover, considering its great taxonomic diversity, investigation related to the search of new bioactive compounds from the marine environment has seen in almost unlimited field [1, 2].

Marine-based bioactive food ingredients can be derived from many sources, including marine plants, microorganisms, and sponges, all of which contain their own unique set of biomolecules [1]. These natural products have a wide range of therapeutic properties, including antimicrobial, antioxidant, antihypertensive, anti-coagulant, anticancer, anti-inflammatory, wound-healing and immune modulator, and other medicinal effects [3]. The new cembranoids, crassumolides A and C, from the soft coral *Lobophytum crassum* inhibited the expression of iNOS and COX-2 [4]. A new briarane-type diterpenoids frajunolides B and C, isolated from the Taiwanese gorgonian *Junceella fragilis*, significantly inhibited superoxide anion and elastase generation from human neutrophils in vitro [5]. A novel compound carteramine A in the marine sponge *Stylissa carteri* inhibited neutrophil chemotaxis [6].

4.1.1 Plant Sources

Moreover, novel bioactive peptides from plants, animals, marine sponges, bacterium, and microalgae are also described with their pharmacological effects in relation with anti-inflammation. Plant source include a variety of medicinal plants used in herbalism and treatment of disease. Moreover, most of the plants are considered as important source of anti-inflammation, and as a result of that, these plants are recommended for their therapeutic values (Table 4.1).

Aloe vera is a succulent plant species. *Aloe vera* is a stemless or very short-stemmed succulent plant growing to 60–100 cm tall. The leaves are thick and fleshy, green to gray green, and with some varieties showing white flecks on their upper and lower stem surfaces. The margin of the leaf is serrated and has small white

Table 4.1 Anti-inflammatory plant sources (terrestrial)

Source	Plant parts	Active constituents	Other medicinal uses	References
Acacia victoriae	Aerial parts	Avicins D and G	Anticancer activity	[7]
Achillea schischkinii	Aerial parts	1,8-Cineole	Antimicrobial and antinociceptive activities	[8, 9]
Adansonia digitata	Fruit pulp	Protein	Antioxidant activity	[10]
Aesculus chinensis	Seeds	Escin Ia, Ib and isoescin Ia,Ib	Antiviral activity	[11, 12]
Agaricus bisporus	Mushroom	Linoleic acid, linolenic acid	Anti-aromatase activity	[13, 14]
Agrocybe aegerita	Mushroom	ABTS, 2,2′-azinobis-(3-ethylbenzthiazoline-6-sulphonic acid)	Antioxidant, anticancer activity	[15–17]
Ajuga decumbens	Whole plant	Ajugaside	Cancer chemopreventive activity	[18, 19]
Allium sativum	Garlic cloves	Allicin	Anticancer effects	[20]
Aloe barbadensis Miller.	Leaves	7-Methoxy-5-methylchromone	Antimicrobial activity	[21–23]
Alpinia blepharocalyx	Seeds	Diarylheptanoids	Antiproliferative activity	[24, 25]
Annona squamosa	Stems	Annomosin A, annosquamosin C, D, F, and G	Antidiabetic, antioxidant activities	[26, 27]
Aristolochia mollissima	Dried roots and stems	Mandolins R, S, U, W, and X	Antimicrobial, cytotoxicity activity	[28, 29]
Arnebia euchroma	Root	Alkannin, shikonin	Antitumor activity	[30]
Artocarpus altilis	Buds	Cycloaltilisin	Anticonvulsant and antibacterial activity	[31–33]
Artocarpus communis and A. heterophyllus	Leaves	Cycloartomunin	Antioxidant and antibacterial activities	[34–36]
Asparagus officinalis	Aerial parts	Asparagusic acid	Cytotoxic activity	[37, 38]
Astragalus oleifolius	Root	Macrophyllosaponins A–D	Leishmanicidal activity	[39, 40]
Atractylodes lancea	Rhizomes	Atractylochromene	Antimicrobial activity	[41, 42]
Atractylodes macrocephala	Root	Biatractylolide	Aromatase inhibiting activities	[43, 44]
Azima tetracantha	Roots and seeds	Rhamnetin	Antipyretic activity	[45, 46]
Baccharis trimera	Aerial parts	Genkwanin, cirsimaritin, hispidulin	Antimutagen, antihepatotoxic activity	[47, 48]

(continued)

Table 4.1 (continued)

Source	Plant parts	Active constituents	Other medicinal uses	References
Bacopa monnieri	Whole plant	Bacopasides VI–VIII	Antihyperglycemic activity	[49, 50]
Bauhinia tarapotensis	Leaves	Glucopyranoside, isoacteoside, indole-3-carboxylic acid	Antioxidant activity	[51]
Beesia calthaefolia	Whole plants	Beesiosides A–F	Antipyretic, analgesic, antioncotic activities	[52]
Benincasa hispida	Fruit	β-(1→4)-D-galactan	Antioxidant activity	[53, 54]
Bidens leucantha	Seeds	7-O-Glycosides 1–4	Anti-HIV activity	[55]
Biophytum petersianum	Aerial parts	Galacturonic acid and rhamnose	Immunomodulating activity	[56, 57]
Brosimum acutifolium	Bark	Acutifolins B–F	Cytotoxic activity	[58, 59]
Brucea antidysenterica	Stem	Bruceantin	Amoebicidal activity	[60, 61]
Buddleja globosa	Flower, stem, and root	Buddledone	Antihepatotoxic activity	[62, 63]
Bursera suntui and Bursera kerberi	Stems	(1S,3Z,7E,11S,12S)-(±)-Verticilla-3,7-dien-12,20-diol and (1S,3Z,7E,11S,12S)-(±)-verticilla-3,7-dien-12,20-diol 20-acetate	Anticancer activity	[64, 65]
Caesalpinia crista	Seed kernels	Caesalpinins MA–ME, norcaesalpinins MA–MC	Antimalarial activity	[66, 67]
Camellia oleifera	Seed	2,5-Bis-benzo[1,3]dioxol-5-yl-tetrahydro-furo [3,4-d][1,3]dioxin	Antioxidant activity	[68]
Camellia sinensis	Leaves	Theaflavin, theaflavin 3,3'-digallate, catechins	Antioxidant and anticancer activity	[69–71]
Cassia spectabilis	Flowers	(−)-3-O-Acetylspectaline	Anti-*Candida albicans* biofilm activity	[72, 73]
Cassia tora	Leaves	α-Tocopherol	Antioxidant activity	[74]
Carpesium longifolium	Aerial parts	Xanthanolides	Cytotoxic activity	[75, 76]
Caulophyllum thalictroides	Rhizomes	Taspine and magnoflorine	Cytotoxic activity	[77]
Cedrelopsis microfoliata	Stem bark	Microfolione	Superoxide-scavenging activity	[78]

Celastrus orbiculatus	Aerial parts	(−)-Epicatechin and (−)-epiafzelechin	Antinociceptive and hypnotic activity	[79, 80]
Chenopodium quinoa	Seeds	Phytolaccagenic acid	Antioxidant, anticancer activity	[81]
Cinnamomum osmophloeum	Leaves	1,8-Cineole	Antiproliferation tumoral cell activity	[82]
Citrus junos Sieb	Fruit	Vitamin C, hesperidin	Antioxidant activity	[83]
Citrus reticulata	Fruit peels	Nobiletin and tangeretin	Antibacterial activity	[84, 85]
Citrus sinensis	Peel	Polymethoxyflavones	Anticarcinogenic, antiatherogenic, and radical-scavenging activity	[86–88]
Coix lachrymajobi	Hulls	Coniferyl alcohol, syringic acid, ferulic acid	Antioxidant activity	[89]
Conyza blinii	Aerial parts	Conyzasaponins A, B, C, and G	Expectorant, antitussive, antibacterial activities	[90]
Coptis japonica	Roots and rhizomes	Berberine sulfate, berberine iodide, palmatine chloride, and woorenosides	Anti-photooxidative activity	[91–93]
Crinum yemense	Bulbs	Yemenines A, B, and C	Tyrosinase inhibitory activity	[94, 95]
Croton eluteria	Bark	Eluterins A–J	Anticancer, antioxidant activity	[96, 97]
Croton tonkinensis	Leaves	Kaurane diterpenoids, ent-kaurane-type diterpenoids	Anti-staphylococcal activity	[98, 99]
Cryptolepis buchanani	Latex	Cryptolepain	Anti-dermatophyte activity	[100, 101]
Cryptolepis sanguinolenta	Root	Cryptolepine	Antihyperglycemic properties	[102]
Curcuma domestica and Curcuma xanthorrhiza	Rhizomes	Curcuminoids	Lipid-reducing and sedative actions	[103, 104]
Curcuma longa	Rhizome	Curcuminoids	Antioxidative, anticarcinogenic, activities	[105, 106]
Cyclanthera pedata Schrab	Seeds	Cucurbitacin glycosides	Antioxidant activity	[107]
Cymbopogon citratus	Leaves	Isoorientin, isoscoparin, swertiajaponin	Antioxidant activity	[108]
Daphniphyllum calycinum	Seeds	Daphcalycinosidine C, yuzurimine E and yuzurimic acid B	Antioxidant activity	[109, 110]

(continued)

Table 4.1 (continued)

Source	Plant parts	Active constituents	Other medicinal uses	References
Dendrobium moniliforme	Stems	Dendrosides A and C, moniliformin, denbinobin	Antioxidation activity, immunomodulatory activity	[111–113]
Deprea subtriflora	Whole plant	Subtrifloralactones A–J, C-18 norwithanolides	Chemopreventive activity	[114]
Digitalis purpurea	Leaves	Digitoxin	Cardiovascular activity	[115]
Dimocarpus longan	Flower	Tetracosanoic acid	Antioxidative, antiapoptotic activity	[116, 117]
Dysoxylum binectariferum	Aerial part	Rohitukine and forskolin	Immunomodulatory, anticancer, intraocular pressure-lowering, and cardiotonic properties	[118]
Echinacea purpurea, E. angustifolia, E. pallida	Roots	Cichoric acid, echinacoside	Immunomodulatory activity	[119]
Echinochloa utilis	Grass	Serotonin, luteolin and tricin	Antioxidant activity	[120]
Emblica officinalis	Fruits	Emblicanin A, emblicanin B	Hepatoprotective activity	[121–124]
Epilobium angustifolium	Aerial parts	Flavonoids	Antioxidant activity	[125]
Eryngium alpinum	Roots	Glucopyranosyl rosmarinic acid	Antioxidant activity	[126]
Ervatamia heyneana	Latex	Ervatamins B and C	Antitumor activity	[127, 128]
Erythroxylum pervillei	Roots	Tropane alkaloid	Anticancer activity	[129, 130]
Erythrina variegata	Roots	Abyssinone V and erycrystagallin	Antibacterial activity	[131, 132]
Ficus septica	Leaves	Ficuseptines B–D	Anticancer, immunomodulatory activity	[133, 134]
Fomitopsis pinicola	Fruit	Lanostane, triterpenoids	Antioxidant activity	[135, 136]
Fragaria ananassa	Fruit	p-Coumaroylglucose	Antioxidant activity	[137]
Ganoderma colossum, Ganoderma lucidum	Fruit	Colossolactones	Antibacterial activity	[138, 139]
Garcinia indica	Fruit	Garcinol	Anticancer, antioxidative, and anti-glycation activities	[140–142]
Garcinia mangostana	Leaves	Garcimangosones A, B, C, D	Antimycobacterial activity	[143–145]

Garcinia multiflora	Stems	Garcinianones A and B	Anti-HIV activity	[113, 146]
Gardenia jasminoides	Fruit	Ursolic acid and genipin	Antioxidant activities and inhibitory effects on the growth of *Helicobacter pylori*	[147]
Garcinia Subelliptica	Heartwood, seed	Garcinielliptone P and hyperellipton HF	Superoxide-scavenging activity	[148, 149]
Geranium niveum	Aerial parts	Geranins	Antiviral, antibacterial, enzyme-inhibiting, antioxidant, and radical-scavenging properties	[150, 151]
Glycyrrhiza glabra	Roots, rhizomes, and leaves	Diisopentenylstilbene and isopentenylstilbene	Anticonvulsant activity	[152, 153]
Glycyrrhiza uralensis	Root	Glycyrrhizin	Antioxidant	[154, 155]
Gnetum cleistostachyum	Aerial parts	Gnetucleistol F, gnetofuran A, lehmbachol D, gnetifolin F, and gnetumontanin C	Anticarcinogenic activity	[156–158]
Harpagophytum procumbens	Root	Harpagide	Anticonvulsant activity	[159, 160]
Hedoytis diffusa	Whole plants	Galactopyranoside and O-acylated iridoid glycosides	Anticancer activity	[161, 162]
Heisteria acuminata	Bark	Pentadeca-6,8,10-triynoic acid	–	[163]
Helianthus annuus	Seed	Terpenoids	Antioxidant activity	[164]
Helichrysum italicum	Aerial parts	Acetophenone glucosides	Antioxidant	[165, 166]
Heterotheca inuloides	Aerial parts	Cadalen-15-oic acid	Antioxidant activity	[167, 168]
Hedoytis diffusa	Whole plants	Galactopyranoside and O-acylated iridoid glycosides	Anticancer activity	[161, 162]
Humulus lupulus	Hops (female flower)	Prenylflavonoids xanthohumol,	Neuropharmacological activity	[169, 170]
Hymenaea palustris	Leaves	Flavonoids	Antimicrobial activity	[171]
Idesia polycarpa	Fruits	Idesolide	Antioxidant activity	[172–174]

(continued)

Table 4.1 (continued)

Source	Plant parts	Active constituents	Other medicinal uses	References
Ilex paraguariensis	Leaves	Caffeine, theobromine, phytol, vitamin E, squalene, and stigmasterol	Antimicrobial activity	[175, 176]
Inula viscosa	Leaves	Rhamnocitrin, 3β-hydroxyilicic acid, and 2α-hydroxyilicic acid	Antifungal activity and wound-healing properties	[177–181]
Ipomoea arborescens	Roots	Arboresins and murucins	Cytotoxicity property	[182]
Ipomoea batata	Roots	ß-Carotene	Antioxidant and anti-atherosclerotic activity	[183–185]
Ipomoea pes-caprae	Aerial parts	Jalapinolic acid, pescaproside A and pescapreins I–IV, stoloniferin III	Antioxidant activity	[186, 187]
Iryanthera juruensis	Leaves	Sargachromenol, sargaquinoic acid, juruenolic acid	Antioxidant activity	[188, 189]
Iryanthera megistophylla	Stem bark	Megislignan and megislactone	Antibacterial, antifungal, antiviral, and antiacetylcholinesterase activities	[190, 191]
Isodon adenantha	Aerial parts	Adenanthins B–L,	Immunoregulatory activity	[192]
Isodon excisa	Leaves	Excisanins	Antitumor and antibacterial activity	[193]
Isodon xerophilus	Leaves	Xerophilusins A–C, macrocalin B	Cytotoxic activity	[194, 195]
Ixeris chinensis	Whole plant	Chinensiolides D and E	Anticancer activity	[196, 197]
Justicia ciliata	Whole plant	Ciliatoside	Cytotoxic activity	[198–200]
Justicia hyssopifolia	Aerial parts	Elenoside	Cytotoxic activity	[200, 201]
Laggera pterodonta	Aerial parts	Eudesmane sesquiterpenes	Immunosuppressant activity	[202, 203]
Leptadenia pyrotechnica	Whole plant	Pregnane glycosides	Antiproliferative activity	[204]
Lippia multiflora	Leaves	Phenylpropanoid glycosides	Antioxidant activity	[205]
Lithospermum erythrorhizon	Roots	Shikonin	Antimutagenic activity	[206]
Machaerium multiflorum	Whole plant	Machaeriol C and D, machaeridiol A, B, and C	Antimicrobial and antiparasitic activity	[207]
Maesa lanceolata	Leaves	Maesasaponin mixture B	Antiangiogenic activity	[208, 209]

Mallotus repandus	Stem and root bark	Berginin	Scavenging effects	[210–212]
Malus domestica	Cores and peels	Pectin	Anticarcinogenic and antimetastatic action	[213]
Malus kornicensis	Fruits	Catechin, epicatechin, cyanidin-3-O-β-galactopyranoside, and amygdalin	Antioxidant activity	[214]
Maytenus laevis	Bark	Canophyllol	Antitumor effects	[215]
Melaleuca alternifolia	Leaves	α-Terpinene, α-terpinolene, γ-terpinene	Antioxidant activity	[216]
Mesona procumbens	Leaves	Protocatechuic acid, syringic acid	Antioxidant activity	[217, 218]
Merulius incarnatus	Mushroom	Trienylresorcinol	Antimicrobial activity	[219, 220]
Microtropis fokienensis	Roots	Forkienin and dihydroagarofuranoid sesquiterpenes	Cytotoxic and antitubercular activity	[221]
Microtropis japonica	Stems	Ursane-type triterpenoids, 2,3-seco-oleanane-type triterpenoid	Antitubercular activity	[222]
Mikania thapsoides	Aerial parts	Sesquiterpene lactones	Antiprotozoal, antiviral activities	[223, 224]
Monascus pilosus	Fermented rice	Monascin, ankaflavin, rubropunctatin	Chemopreventive effects	[225, 226]
Montanoa hibiscifolia	Aerial parts	Montabibisciolides and germacrolides	Phagodeterrent activity	[227, 228]
Morinda citrifolia	Fruits	Scopoletin	Antispasmodic and vasodilator activities	[229, 230]
Moringa oleifera and Moringa stenopetala	Seeds, leaves, bark	Benzyl glucosinolate	Antibiotic activity	[231, 232]
Morus nigra and Morus alba	Fruits, bark	Cyanidin 3-O-glucoside, cyanidin 3-O-rutinoside, pelargonidin 3-O-glucoside, pelargonidin 3-O-rutinoside	Antioxidant activity	[233, 234]
Murraya koenigii	Leaves	Mahanimbine	Antimicrobial activity	[235–237]
Myristica cinnamomea	Fruits	Myristinins	Antifungal and anti-quorum-sensing activity	[238, 239]
Myrianthus holstii	Root	*M. holstii* lectin (MHL)	Anti-HIV activity	[240]

(continued)

Table 4.1 (continued)

Source	Plant parts	Active constituents	Other medicinal uses	References
Nelumbo nucifera	Leaves	Linoleic acid, oleic acid, and behenic acid	Antioxidant activity	[241]
Olea europaea	Leaves	β-Carotene, α-tocopherol, oleuropein, and rutin	Antimicrobial activity and antioxidant Activity	[242–244]
Ononis spinosa	Root	Spinonin, ononin, and homopterocarpin	Antifungal activity	[245, 246]
Opuntia ficus indica	Fruits	Ascorbic acid, polyphenols, and flavonoids	Antioxidant, antiulcerogenic activity	[247]
Origanum onites	Aerial parts	Carvacrol	Antimicrobial activity	[248, 249]
Ornithogalum caudatum	Leaves	Homoisoflavanone glycosides	Antitumor activities	[250, 251]
Orthosiphon stamineus	Leaves	Orthosiphols U–Z	Antiapoptotic and antioxidant activity	[252, 253]
Paeonia suffruticosa	Root	Mudanpinoic acid A	Acaricidal activities	[254–256]
Panax ginseng	Root	p-Coumaric	Antioxidant activities	[257, 258]
Panax japonicus	Root	Panajaponol and pseudoginsenoside RT1 butyl ester	Antiallergic activity	[259, 260]
Perilla frutescens	Leaves and seeds	Catechin, ferulic acid, apigenin, luteolin, rosmarinic acid	Antioxidant and antimicrobial activity	[261–264]
Peucedanum ostruthium	Rhizomes	Coumarins, oxypeucedanin ostruthol, imperatorin	Antiphlogistic and antipyretic activities	[265, 266]
Physalis peruviana	Whole plant	Dienolide	Antioxidant activities	[267, 268]
Phyllanthus emblica	Roots	Phyllaembic acids B and C and phyllaemblicin D	Antioxidant activity	[269, 270]
Picrorhiza kurroa	Root	Picroside II	Antioxidant and antineoplastic activities	[271, 272]
Picrorhiza scrophulariaeflora	Dried rhizomes	Picracin and deacetylpicracin	Immunomodulatory activity	[273, 274]
Pinus mugo	Leaves	Δ³-Carene, camphene, α-pinene,	Antioxidant, antibacterial activity	[275]
Piper nigrum	Roots	Nigramides A–O, nigramides P–S	Larvicidal activity	[276, 277]

Pistacia vera	Gum	α-Pinene, β-pinene, trans-verbenol, camphene, trans-pinocarveol, and limonene	Antimicrobial activity	[278]
Plantago major	Seeds and leaves	Aucubin	Antiviral activity	[279]
Polygonatum odoratum	Roots	Homoisoflavanones	Antioxidant activities	[280, 281]
Prunus amygdalus	Fruit	3-Prenyl-4-O-β-d-glucopyranosyloxy-4-hydroxylbenzoic acid	Antioxidant activity	[282]
Prunus cerasus	Fruit	Anthocyanins and cyanidin	Antioxidant	[283]
Prunus dulcis	Nut	Chlorogenic acid,	Antioxidant activity	[284]
Pseudocydonia sinensis	Fruit	(−)-Epicatechin	Antioxidant and anti-influenza activity	[285]
Pterodon polygalaeflorus	Seeds	Vouacapane	Larvicidal activity	[286, 287]
Pulsatilla cernua	Root	3,4-Dihydroxycinnamic acid and 4-hydroxy-3-methoxycinnamic acid	Selective growth inhibitor	[288, 289]
Punica granatum	Fruits	Ellagitannins and punicalagin	Antiplasmodial and antioxidant activity	[290, 291]
Rheum emodi	Roots	Emodin 8-O-β-d-glucopyranosyl-6-O-sulfate	Antioxidant activity	[292]
Rhododendron dauricum	Leaves	Azelain	Antiulcerogenic activity	[293]
Ribes Species (R. nigrum, R. rubrum, R. grossularia)	Seed	Tocopherol	Anti-influenza activity	[294, 295]
Rosa canina	Fruits	Galactolipid-3-O-β-d-galactopyranosyl glycerol	Antioxidant effect	[296–299]
Rosmarinus officinalis	Leaves	Rosmarinic acid, ursolic acid, oleanolic acid, and micromeric acid	Spasmolytic activity and free radical-scavenging activity	[300–302]
Rubus pungens	Aerial parts	Rubupungenosides A and B	Antibacterial activity	[303–305]
Ruscus aculeatus	Fruit	Pelargonidin 3-O-rutinoside, pelargonidin 3-O-glucoside, and pelargonidin 3-O-trans-p-coumarylglucoside	Anti-elastase and anti-hyaluronidase activities	[306, 307]

(continued)

Table 4.1 (continued)

Source	Plant parts	Active constituents	Other medicinal uses	References
Russula griseocarnosa	Mushroom	Phenolics, flavonoids, ergosterol, and β-carotene	Antioxidant activities	[308]
Salacia chinensis	Stems	Salasones A, B, C and salaquinone A	Antihyperlipidemic activity	[309, 310]
Salvia broussonetii	Root	Brussonol and iguestol	Insecticidal activity	[311]
Salvia officinalis	Aerial parts	1,8-Cineole, (−)-thujone, (±)-	Antibacterial, allelopathic, antioxidant activities	[312, 313]
Salvia miltiorrhiza	Root	Salvianen	Antioxidant activities	[314, 315]
Salvia trijuga	Root	Trijugins A–I	Cytotoxicity activity	[316]
Santolina insularis	Aerial parts	Sesquiterpenoids and the trans-chrysanthemyl monoterpenoid	Antiviral activity	[317, 318]
Santolina oblongifolia	Flower tops	Hemiarin, aesculetin, scopolin, and scopoletin	Antiviral activity	[319]
Sarcophyte piriei	Rhizome	Diinsinin and naringenin	Oxidative stress protection	[320, 321]
Satureja parnassica	Aerial parts	Carvacrol	Bactericidal properties	[322]
Saussurea lappa	Root	Lappadilactone, dehydrocostuslactone, and costunolide	Antiarthritic activity	[323, 324]
Scutellaria baicalensis	Root	Baicalein, baicalin	Anticancer activity	[325–327]
Sempervivum tectorum	Leaves	5-Diethoxyphosphoryl	Antioxidant activity	[328]
Sesamum Indicum	Seeds	Vanillic acid	Antioxidant efficiency, antinitrosating properties	[329]
Smilax medica	Roots	Saponins	Antifungal activity	[330]
Spinacia oleracea	Leaves	Phenolic	Antioxidant activity	[331]
Stereospermum personatum	Stem and stem bark	Sterequinones F–H	Antioxidant activity	[332]
Syzygium aromaticum	Flower bud	Phenylpropanoids	Antimutagenic activity	[127]

Species	Plant part	Compound	Activity	References
Symphytum asperum	Roots	Poly[3-(3,4-dihydroxyphenyl)glyceric acid]	Antioxidant, anti-lipoperoxidant, and cytotoxic properties	[333, 334]
Tabebuia impetiginosa	Bark	Menadione and plumbagin	Antibacterial activity	[335, 336]
Tadehagi triquetrum	Leaves	Triquetrumones A, B, and C	Anthelmintic activity	[337]
Tagetes lucida	Aerial parts	Coumarins(7,8-dihydroxycoumarin, umbelliferone)	Antifungal and antibacterial activities	[338]
Tamarindus indica	Fruit, leaves, flowers	Vitamin C and β-carotene	Antioxidant, antimicrobial, and antifungal activities	[322]
Tanacetum microphyllum	Aerial part	Santin, ermanin	Antiulcerogenic activity	[339–341]
Tanacetum parthenium	Whole plant	Parthenolide	Anticancer activity	[342–344]
Tetraclinis articulata	Leaves and wood	Pimaranes	Antioxidant activity	[345, 346]
Tinospora cordifolia	Stem	Tinocordiside	Antioxidant and immunomodulatory activity	[347–349]
Tournefortia sarmentosa	Stems	Tournefolins A, B, and C	Anti-lipid-peroxidative activity	[350, 351]
Uncaria sinensis	Fruit	Ursolic acid	Antimutagenic action	[352–354]
Uncaria tomentosa	Bark	Tomentosides A and B	Antiproliferative activity	[355, 356]
Vaccinium arctostaphylos	Leaves and fruits	Gentisic, salicylic, and trans-cinnamic acids	Antioxidative, antitumor, antiviral, vasoprotective, and antifungal activities	[357, 358]
Vaccinium ashei	Fruit	Flavonoids	Antioxidant activity	[359, 360]
Vaccinium macrocarpon	Fruit	Phenylboronic acid	Antioxidant, antitumor activity	[361–363]
Vaccinium myrtillus	Fruit	Anthocyanins	Antimicrobial activity	[364]
Vaccinium vitis-idaea and Vaccinium microcarpon	Fruit	α-Amyrin, β-amyrin, and proanthocyanidins	Antioxidant, antimicrobial, antiadhesive, properties	[365–367]
Vanillosmopsis pohlii	Heartwood and leaves	α-Bisabolol, β-pinene, E-caryophyllene	Insecticidal activity	[368]
Vatica oblongifolia	Stem bark	Hopeaphenol A and isohopeaphenol A	Antimicrobial activity	[369, 370]
Visnea mocanera	Leaves and fruits	Procyanidins	Antimicrobial, analgesic, antiulcerogenic, hemostatic, astringent, cicatrizant, vulnerary, and psychostimulant activities	[371]

(continued)

Table 4.1 (continued)

Source	Plant parts	Active constituents	Other medicinal uses	References
Vitis vinifera	Seed	Rutin	Antioxidant activity	[372–374]
Withania somnifera	Seeds	Withanolides	Cytotoxic activity	[375, 376]
Yucca schidigera	Bark	Yuccaols D and E	Antioxidant activity	[377]
Zanha africana	Root bark	Saponins, zanhasaponins, and cyclitol pinitol	Cytotoxic activity	[378, 379]
Zingiber cassumunar	Rhizomes	Curcuminoids, cassumunin	Antioxidant activity	[380]
Ziziphus jujuba	Leaves	Ziziphin	Prevention of atherosclerosis	[381–383]

teeth. The flowers are produced in summer on a spike up to 90 cm tall, each flower being pendulous, with a yellow tubular corolla 2–3 cm long. Extracts from *Aloe vera* are widely used in the cosmetics and alternative medicine industries, being marketed as variously having rejuvenating, healing, or soothing properties. Researchers found that oral administration of aloe vera might be effective in reducing blood glucose in diabetic patients and in lowering blood lipid levels in hyperlipidemia. The topical application of aloe vera does not seem to prevent radiation-induced skin damage. It might be useful as a treatment for genital herpes and psoriasis. The evidence regarding wound healing is contradictory. Topical application of aloe vera may also be effective for genital herpes and psoriasis. There is, however, little scientific evidence of the effectiveness or safety of *Aloe vera* extracts for either cosmetic or medicinal purposes, and what positive evidence is available is frequently contradicted by other studies. Like other *Aloe* species, *Aloe vera* forms arbuscular mycorrhiza, a symbiosis that allows the plant better access to mineral nutrients in soil (Fig. 4.1a) [384–388]

Rosa canina commonly known as the dog rose is a variable climbing wild rose species. It is native to Europe, northwest Africa, and western Asia. It is a deciduous shrub normally ranging in height from 1 to 5 m, though sometimes it can scramble higher into the crowns of taller trees. Its stems are covered with small, sharp, hooked prickles, which aid it in climbing. The leaves are pinnate, with 5–7 leaflets. The flowers are usually pale pink, but can vary between a deep pink and white. They are 4–6 cm diameter with five petals and mature into an oval 1.5–2 cm red-orange fruit or hip (Fig. 4.1b).

Rosmarinus officinalis, commonly known as rosemary, is a woody, perennial herb with fragrant, evergreen, needle-like leaves and white, pink, purple, or blue flowers, native to the Mediterranean region. It is a member of the mint family Lamiaceae, which includes many other herbs. Rosemary is an aromatic evergreen

Fig. 4.1a *Aloe vera*

Fig. 4.1b *Rosa canina*

Fig. 4.1c *Rosmarinus officinalis*

shrub that has leaves similar to hemlock needles. The leaves are used as a flavoring in foods such as stuffings and roast lamb, pork, chicken, and turkey. It can withstand droughts, surviving a severe lack of water for lengthy periods. Forms range from upright to trailing; the upright forms can reach 1.5 m tall, rarely 2 m. The leaves are evergreen, 2–4 cm long and 2–5 mm broad, green above, and white below, with dense, short, woolly hair (Fig. 4.1c) [389, 390].

Prunus cerasus also called sour cherry, tart cherry, or wild cherry is a species of *Prunus* in the subgenus *Cerasus* (cherries), native to much of Europe and south-western Asia. It is closely related to the sweet cherry (*Prunus avium*), but has a fruit

Fig. 4.1d   *Prunus cerasus*

that is more acidic, has greater nutritional benefits, and may have greater medicinal effects. It has twiggy branches, and its crimson-to-near-black cherries are borne upon shorter stalks. There are two varieties of the sour cherry: the dark-red morello cherry and the lighter-red amarelle cherry. Medicinally, sour cherries may be useful in alleviating sleep problems due to its high melatonin content, a compound critical in regulating the sleep-wake cycle in humans. Further research is going on exploring the significant benefit of sour cherries in several medical applications like inflammation, pain management and others (Fig. 4.1d) [391, 392].

Cinnamomum osmophloeum, commonly known as pseudocinnamomum or indigenous cinnamon, is a medium-sized evergreen tree in the genus *Cinnamomum*. It is native to broad-leaved forests of central and northern Taiwan. Cinnamaldehyde, an essential oil extracted from C. osmophloeum, has numerous commercial uses. Also, it is a xanthine oxidase inhibitor, hence a potential drug for treatment of hyperuricemia and related medical conditions including gout (Fig. 4.1e) [393].

Inula viscose (L.) is a bushy plant, usually evergreen, belonging to the family of the Asteraceae which is quite common in the Mediterranean regions. It is a woody shrub and abundantly branched, vigorous, with erect branches, typically 50–80 cm tall, exceptionally up to 150 cm, and with pubescent leaves and shoots, glandulous and stickier, emanating a strong odour of aromatic resin. The leaves are alternate or irregularly scattered, the lower ones sessile, the upper ones amplessicauli, with entire margin and lamina lanceolate, toothed, or serrated teeth sparse. They are usually persistent. The flowers are grouped in showy yellow flower heads 1–1.5 cm in diameter, in turn, gathered in abundant and long terminal panicles. Peripheral flowers are feminine and ligulate with long, straight, golden yellow color; the internal ones are hermaphrodites, with tubular corolla, ending with five golden yellow teeth. The fruit is an achene about 2 mm long, equipped with a hairy pappus hairs gathered at the base (Fig. 4.1f) [394, 395].

Fig. 4.1e *Cinnamomum osmophloeum*

Fig. 4.1f *Inula viscose*

Aesculus chinensis is also known as Chinese horse chestnut or Chinese buckeye. It is a rounded deciduous tree and member of Hippocastanaceae family. *A. chinensis* is native to China. The tree can grow to a height of 25 m and up to 10 m wide. *A. chinensis* is used in the treatment of stomach aches. The seed is antirheumatic and emetic in action and needs to be leached of toxins before it becomes safe to eat (Fig. 4.1g) [396].

Coptis japonica, also called Goldthread or Canker Root, is a species of flowering plants in the family Ranunculaceae, native to Asia and North America. It is used as a medicinal herb in China and as a bitter tonic for dyspepsia in the Himalayan

Fig. 4.1g *Aesculus chinensis*

Fig. 4.1h *Coptis japonica*

regions of India. It is also believed to help insomnia in Chinese herbology. The roots contain the bitter alkaloid berberine. The dried roots were commercially marketed in Canada onto areas affected by thrush (candidiasis) infection (Fig. 4.1h) [397].

Heterotheca inuloides is a member of the family Asteraceae. They are annual or perennial herbs, usually with primary roots, sometimes with rhizomes, and stem base sometimes slightly woody. Stems are erect or ascending, usually several with numerous ascending branches above the midpoint, sometimes only a few-branched towards the tip. Basal leaves are often withered or absent at the time of flowering;

Fig. 4.1i   *Heterotheca inuloides*

the leaf is narrowly to broadly oblanceolate, tapered at the base with a winged petiole, the margins entire, slightly wavy, or toothed in various ways. Stem leaves are
slightly moderately reduced to the root tip, sessile, and narrowly oblanceolate to
oblong-lanceolate and oblong-ovate, with the margins entire or variously toothed
and surfaces and especially margins moderately to densely hairy, sometimes glandular (Fig. 4.1i) [398].

Artocarpus is a genus of approximately 60 trees and shrubs of Southeast Asian
and Pacific origin, belonging to the mulberry family, Moraceae. All *Artocarpus* species are laticiferous trees or shrubs that are composed of leaves, twigs, and stems
capable of producing a milky sap. The fauna type is monoecious and produces uni-
sexual flowers; furthermore, both sexes are present within the same plant. The plants
produce small, greenish, female flowers that grow on short, fleshy spikes. Following
pollination, the flowers grow into a syncarpous fruit, and these are capable of growing into very large sizes. The stipulated leaves vary from small and entire (*Artocarpus
integer*) to large and lobed (*Artocarpus altilis*), with the cordate leaves of the species *A. altilis* ending in long, sharp tips (Fig. 4.1j) [399].

Santolina is a genus of flowering plants in the family Asteraceae, native to the
Mediterranean. They are small evergreen shrubs growing 10–60 cm tall. The leaves
are simple and minute in some species, or pinnate, finely divided in other species,
often densely silvery hairy, and usually aromatic. The composite flower heads are
yellow or white, produced in dense globose capitulae 1–2 cm in diameter, and on
top of slender stems held 10–25 cm above the foliage. *Santolina* species are used as
food plants by the larvae of some Lepidoptera species (Fig. 4.1k).

Vitis vinifera belongs to the family Vitaceae. It is native to the Mediterranean
region, central Europe, and southwestern Asia. A grape is a fruiting berry of the
deciduous woody vines of the botanical genus *Vitis*. Grapes can be eaten raw or they
can be used for making wine, jam, juice, jelly, etc. Grapes are non-climatic type of
fruit, generally occurring in clusters. Grape seed extracts are industrial derivatives

Fig. 4.1j *Artocarpus altilis*

Fig. 4.1k *Santolina insularis*

from whole grape seeds that have a great concentration of vitamin E, flavonoids, linoleic acid, and phenolic procyanidins. The typical commercial opportunity of extracting grape seed constituents has been for chemicals known as polyphenols having antioxidant activity in vitro (Fig. 4.1l).

Picrorhiza scrophulariaeflora belongs to the family Scrophulariaceae. The root contains several glycosides and active constituents including iridoid glycosides, amphicoside, catalpol, aucubin, androsin, and cucurbitacin glycosides. It is stimulant, expectorant, powerful diaphoretic. It is used in hair tonics to stimulate hair growth. The leaf contains a parasympathetic stimulant pilocarpine (0.5 %). It is an obsolete medicinal herb, but is used in the production of pilocarpine (Fig. 4.1m).

Fig. 4.1l *Vitis vinifera*

Fig. 4.1m *Picrorhiza scrophulariaeflora*

Helichrysum italicum is a flowering plant of the daisy family Asteraceae. It is sometimes called the curry plant because of the strong smell of its leaves. It grows on dry, rocky, or sandy ground around the Mediterranean. The stems are woody at the base and can reach 60 cm or more in height. The clusters of yellow flowers are produced in summer; they retain their color after picking and are used in dried flower arrangements.

Fig. 4.1n *Helichrysum italicum*

The plant produces oil from its blossoms which is used for medicinal purposes. It is anti-inflammatory, fungicidal, and astringent. It soothes burns and raw chapped skin. It is used as a fixative in perfumes and has an intense fragrance. This plant is sometimes used as a spice. Rather, it has a resinous, somewhat bitter aroma reminiscent of sage or wormwood and is used like these: the young shoots and leaves are stewed in Mediterranean meat, fish, or vegetable dishes till they have imparted their flavor and removed before serving (Fig. 4.1n) [400].

Garcinia indica, a plant in the mangosteen family (Clusiaceae), commonly known as kokum, is a fruit-bearing tree that has culinary, pharmaceutical, and industrial uses. It is indigenous to the Western Ghats region of India located along the western coast of the country. Of the 35 species found in India, 17 are endemic. Of these, seven are endemic to the Western Ghats, six in the Andaman and Nicobar Islands, and four in the northeastern region of India. *Garcinia indica* is found in forest lands, riversides, and wastelands. These plants prefer evergreen forests, but sometimes they also thrive in areas with relatively low rainfall. It is also cultivated on a small scale. It does not require irrigation and spraying of pesticides or fertilizers (Fig. 4.1o) [401].

Rubus pungens is a deciduous shrub growing to 3 m. It belongs to the family Rosaceae. It is easily grown in a good well-drained loamy soil in sun or semi-shade. Plants in this genus are notably susceptible to honey fungus. *R. pungens* is suitable for nearly all types of soil and tolerate any pH condition of soil (Fig. 4.1p).

Uncaria is a genus of flowering plants in the family Rubiaceae. It has about 40 species. Their distribution is pantropical, with most species native to tropical Asia, three from Africa and the Mediterranean and two from the neotropics. They are known colloquially as Gambier or cat's claw. The genus name is derived from the Latin word uncus, meaning "a hook." It refers to the hooks, formed from reduced

Fig. 4.1o *Garcinia indica*

Fig. 4.1p *Rubus pungens*

branches that *Uncaria* vines use to cling to other vegetation. Cat's claw (*U. tomentosa*) and the Chinese species are used medicinally. The glycosidic compounds have recognized anti-inflammatory properties, while the alkaloids increase the reactivity of lymphocytes, granting higher response to viral infection. Cat's claw has two varieties depending on whether the alkaloids have four rings or five. Chinese were using it for tanning and noted that the *Uncaria gambir* made "leather porous and rotten." It is also noted that Chinese would chew it with areca nut. It contains many flavan-3-ols (catechins) which are known to have many medicinal properties and are components of Chinese herbal remedies and certain modern medicines. Research has shown that rhamnose, a chemical extracted from *Uncaria* plants, can actively regenerate skin, making it feel plumper and more elastic (Fig. 4.1q) [402].

Symphytum asperum is a flowering plant of the genus *Symphytum* in the family Boraginaceae. Common names include rough comfrey and prickly comfrey. It is native to Asia, and it is known in Europe and North America as an introduced species and sometimes a weed (Fig. 4.1r).

Glycyrrhiza uralensis, also known as Chinese liquorice, is a flowering plant native to Asia, which is used in traditional Chinese medicine. Liquorice root, or

Fig. 4.1q *Uncaria tomentosa*

Fig. 4.1r *Symphytum asperum*

radix glycyrrhizae, is one of the 50 fundamental herbs used in traditional Chinese medicine, where it has the name gan cao. It is usually collected in spring and autumn, when it is removed from the rootlet and dried in the sun. Liquorice root is most commonly produced in the Shanxi, Gansu, and Xinjiang regions of China. As well as traditional Chinese medicine, liquorice root is used in Greco-Arab and Unani medicines, as well as in the traditional medicines of Mongolia, Japan, Korea, Vietnam, Pakistan, India, and other Asian nations. In India, it is referred to as mulethi. The Greco-Arab (Unani) medicine recommends its oral use after removal of external layer to avoid side effects. People with heart conditions or high blood pressure should avoid ingesting extensive amounts of liquorice, as it can further heighten blood pressure and lead to stroke (Fig. 4.1s).

Morinda citrifolia is a tree in the coffee family, Rubiaceae. Its native range extends through Southeast Asia and Australasia, and the species is now cultivated throughout the tropics and widely naturalized. English common names include great morinda, Indian mulberry, noni, beach mulberry, and cheese fruit. It grows in shady forests, as well as on open rocky or sandy shores. It reaches maturity in about 18 months and then yields between 4 and 8 kg (8.8 and 17.6 lb) of fruit every month throughout the year. The plant bears flowers and fruits all year round. The fruit is a multiple fruit that has a pungent odor when ripening and is hence also known as cheese fruit or even vomit fruit. It is oval in shape and reaches 10–18 cm size. At first green, the fruit turns yellow and then almost white as it ripens. It contains many seeds. It is sometimes called starvation fruit. Despite its strong smell and bitter taste, the fruit is nevertheless eaten as a famine food and, in some Pacific islands, even as a staple food, either raw or cooked. Southeast Asians and Australian Aborigines consume the fruit raw with salt or cook it with curry. It is tolerant of saline soils, drought conditions, and secondary soils. It is therefore found in a wide variety of habitats: volcanic terrains, lava-strewn coasts, and clearings or limestone outcrops,

Fig. 4.1s *Glycyrrhiza uralensis*

Fig. 4.1t *Morinda citrifolia*

as well as in coralline atolls. It can grow up to 9 m tall and has large, simple, dark green, shiny, and deeply veined leaves. The seeds are edible when roasted. *Morinda citrifolia* is especially attractive to weaver ants, which make nests from the leaves of the tree. These ants protect the plant from some plant-parasitic insects. The smell of the fruit also attracts fruit bats, which aid in dispersing the seeds. A type of fruit fly, *Drosophila sechellia*, feeds exclusively on these fruits. *Morinda citrifolia* fruit contains a number of phytochemicals, including ligands, oligo- and polysaccharides, flavonoids, iridoids, fatty acids, scopoletin, catechin, beta-sitosterol, damnacanthal, and alkaloids. Although these substances have been studied for bioactivity, current research is insufficient to conclude anything about their effects on human health (Fig. 4.1t) [403, 404].

Zanha golungensis, velvet-fruited *Zanha*, is a species of fruit plants from Sapindaceae family which can be found in Angola, Kenya, Mozambique, Zimbabwe, and the Democratic Republic of the Congo where it is used as door frame and tool handles. It is also used for flooring and for creating toys, railway sleepers, turnery, furniture, and ship designs. The species is a 12–17 m tall shrub, which have 3–6 pairs of leaflets which are ovate, elliptical, and 8–15 cm by 4–8 cm. The petioles are 1–3 mm long, while the pedicels are around 2.5 mm long. It has 4–6 stamens which are 10 mm long with a cup-shaped disk that is hairy with a diameter of 2 mm. Ovary is absent in male species, while the females bare flowers which turn into 3 cm by 2 cm fruits that are hairy and ellipsoid as well. Just like the fruit, the seed is also ellipsoid, but is 1.5 cm by 1 cm and is yellow (sometimes bright orange) in color (Fig. 4.1u).

Gnetum cleistostachyum belongs to the family Gnetaceae. It is a tropical evergreen tree. Unlike other gymnosperms, they possess vessel elements in the xylem. Some species have been proposed to have been the first plants to be insect pollinated as their fossils occur in association with the extinct pollinating scorpionflies.

Fig. 4.1u   *Zanha golungensis*

Fig. 4.1v   *Gnetum cleistostachyum*

Molecular phylogenies based on nuclear and plastid sequences from most of the species indicate hybridization among some of the Southeast Asian species. Fossil-calibrated molecular clocks suggest that the *Gnetum* lineages now found in Africa, South America, and Southeast Asia are the result of ancient long-distance dispersal across seawater (Fig. 4.1v) [405, 406].

   Vaccinium vitis-idaea is a short evergreen shrub in the heath family Ericaceae. It is native to boreal forest and Arctic tundra throughout the Northern Hemisphere from Eurasia to North America. *Vaccinium vitis-idaea* is most commonly known in English as lingonberry or cowberry. *Vaccinium vitis-idaea* spreads by underground

Fig. 4.1w *Vaccinium vitis-idaea*

stems to form dense clonal colonies. Slender and brittle roots grow from the underground stems. The stems are rounded in cross section and grow from 10 to 40 cm in height. The fruit is a red berry 6–10 mm across, with an acidic taste, ripening in late summer to autumn. In folk medicine, *V. vitis-idaea* has been used as an apéritif, astringent, antihemorrhagic, anti-debilitive, depurative, antiseptic, a diuretic, a tonic for the nervous system, and in various ways to treat breast cancer, diabetes mellitus, rheumatism, and various urogenital conditions. In traditional Austrian medicine, the fruits have been administrated internally as jelly or syrup for treatment of disorders of the gastrointestinal tract, kidneys, and urinary tract and fever (Fig. 4.1w) [407, 408].

Zingiber cassumunar, now thought to be a synonym of *Zingiber montanum*, is a species of plant in the ginger family and is also a relative of galangal. It is called plai in Thailand. The rhizome of variant Roxburgh is used medicinally in massage and even in food in Thailand and somewhat resembles ginger root or galangal. In aromatherapy, plai oil is utilized as an essential oil that is believed to ease pain and inflammation. Study by different researchers found that (E)-1-(3,4-dimethoxyphenyl) but-1-ene, an active ingredient of *Zingiber cassumunar* rhizomes, has analgesic and anti-inflammatory properties [409]. In addition to these, plai oil also exhibits antimicrobial activity against a wide range of Gram-positive and Gram-negative bacteria, dermatophytes, and yeasts [410], and the plant has antifungal properties against pathogenic fungi. The plant also contains the essential oils sabinene 31–48 %, terpineol 4–30 % [411], and apparently unique curcuminoid antioxidants, namely, cassumunarin types A, B, and C (Fig. 4.1x) [412].

Tanacetum microphyllum is a species of flowering plants in the aster family, Asteraceae. It is endemic to the Iberian Peninsula. The plant has been used for centuries in Spanish traditional medicine as an anti-inflammatory and antirheumatic. Compounds isolated from extracts of the plant include santin, ermanin, centaureidin, and hydroxyachilin (Fig. 4.1y).

Fig. 4.1x *Zingiber cassumunar*

Fig. 4.1y *Tanacetum microphyllum*

Celastrus orbiculatus is a woody vine of the Celastraceae family. It is commonly called Oriental bittersweet. Other common names include Chinese bittersweet, Asian bittersweet, round-leaved bittersweet, and Asiatic bittersweet. *Celastrus orbiculatus* is considered to be an invasive species in eastern North America. The defining characteristic of the plant is its vines: they are thin, spindly, and have silver to reddish brown bark. They are generally between 1 and 4 cm in diameter. When *Celastrus orbiculatus* grows by itself, it forms thickets; when it is near a tree or shrub, the vines twist themselves around the trunk. The encircling vines have been

Fig. 4.1z *Celastrus orbiculatus*

known to strangle the host tree to death, also true of the American species. The leaves are round and glossy and 2–12 cm long, have toothed margins, and grow in alternate patterns along the vines. Small green flowers produce distinctive red seeds. The seeds are encased in yellow pods that break open during autumn. All parts of the plant are poisonous (Fig. 4.1z) [413].

Rubus is the genus of the rose family; raspberry is the edible fruit of a multitude of plant species in the genus. Raspberries are perennial with woody stems. Traditionally, raspberries were a midsummer crop. Raspberries need ample sun and water for optimal development. Raspberries thrive in well-drained soil with a pH between 6 and 7 with ample organic matter to assist in retaining water. An individual raspberry weights 3–5 g and is made up of around 100 drupelets, each of which consists of a juicy pulp and a single central seed. A raspberry bush can yield several hundred berries a year. Unlike blackberries and dewberries, a raspberry has a hollow core once it is removed from the receptacle. Raspberries are a rich source of vitamin C, manganese, and dietary fiber. Raspberries contain anthocyanin, pigments, ellagic acid, quercetin, gallic acid, cyanidins, pelargonidins, catechins, kaempferol, and salicylic acid. Animal research indicates presence of antioxidant and antiproliferative effects in raspberries (Fig. 4.1aa) [414–417].

Picrorhiza kurroa is one of the major income-generating non-timber forest products found in the Nepalese Himalayas. It is one of the oldest medicinal plants traded from the Karnali zone known as Kutki. It is a perennial herb and is used as a substitute for Indian gentian. Leaves: 5–15 cm long leaves, almost all at the base, often withered. Rhizomes of the plant are 15–25 cm long and woody. Flowers are about 8 mm long, 5-lobed to the middle, and with much longer stamens (Fig. 4.1ab).

Origanum onites is also known as Cretan oregano or Turkish oregano or pot marjoram; it is a perennial growing to 0.3–0.6 m. The flowers are hermaphrodite and are pollinated by bees. It is noted for attracting wildlife. They are used as a

Fig. 4.1aa *Rubus idaeus*

Fig. 4.1ab *Picrorhiza kurroa*

flavoring for salad dressings and added in the final stages of cooking. Having strong thyme-like aroma, the leaves are used as a substitute for oregano or marjoram, but they are inferior in flavor. Its flavor, however, lasts longer in cooked dishes. Medicinally, it is used as antiseptic, antispasmodic, carminative, cholagogue, diaphoretic, emmenagogue, expectorant, stimulant, stomachic, and mildly tonic. It has antimicrobial activities (Fig. 4.1ac) [418, 419].

Dysoxylum densiflorum is a flowering plant, constituting the mahogany family (Meliaceae). *Dysoxylum densiflorum*, locally known as majegau, is the plant "mascot" or floral emblem of Bali. In India, apart from its economic importance for

Fig. 4.1ac   *Origanum onites*

Fig. 4.1ad   *Dysoxylum densiflorum*

building and furniture making, it is an important ingredient in Ayurvedic medicine as many species have curative qualities taken independently or as an ingredient of a medicinal mixture. Some of the uses in Ayurveda are reported as follows: Wood decoction of *D. malabaricum* is used to cure rheumatism and its oil is used to cure eye and ear diseases and a few species are used to cure inflammation, heart disorder, CNS disorder, and also tumor. In Indian tradition and culture, oil is extracted from the seeds of *Dysoxylum malabaricum*, which has wide beneficial application (Fig. 4.1ad).

Fig. 4.1ae *Visnea mocanera*

Visnea mocanera is a species of plant in the Theaceae family. It is found in Portugal and Spain. It is threatened by habitat loss. One of the major uses of *Visnea mocanera* is its use for treatment of wounds caused from physical workouts. Its healing properties have shown remarkable results for faster healing of damaged cells when applied externally. This helps increase the rate of protein synthesis allowing the muscle to rebuild faster and a bodybuilder to train harder, building more muscle in the long run. Extract of *Visnea mocanera* is used as a dietary supplement; it contains high levels of flavan-3-ols and other various nutritional properties. Therefore, this plant is considered a complete nutrition package that can be very effective in building and maintaining good health (Fig. 4.1ae).

Murraya koenigii is a tropical to subtropical tree in the family Rutaceae, which is native to India and Sri Lanka. Its leaves are used in many dishes in India and neighboring countries. Often used in curries, the leaves are generally called by the name "curry leaves," though they are also translated as "sweet neem leaves" in most Indian languages. It is a small tree, growing 4–6 m tall, with a trunk up to 40 cm diameter. The leaves are pinnate, with 11–21 leaflets, each leaflet 2–4 cm long and 1–2 cm broad. They are highly aromatic. The flowers are small, white, and fragrant. The small black shiny berries are edible, but their seeds are poisonous. The leaves of *Murraya koenigii* are also used as an herb in Ayurvedic medicine. They are believed to possess antidiabetic properties (Fig. 4.1af) [420].

Curcuma longa is a rhizomatous herbaceous perennial plant of the ginger family, Zingiberaceae. It is native to tropical Tamil Nadu, in southeast India. Its active ingredient is curcumin, and it has a distinctly earthy, slightly bitter, slightly hot peppery flavor and a mustardy smell. Curcumin has been a center of attraction for potential treatment of an array of diseases, including cancer, Alzheimer's disease, diabetes, allergies, arthritis, and other chronic illnesses. Turmeric is a perennial herbaceous plant, which reaches a stature of up to 1 m. There are highly branched,

Fig. 4.1af *Murraya koenigii*

Fig. 4.1ag *Curcuma longa*

yellow to orange, cylindrical, aromatic rhizomes. The most important chemical components of turmeric are a group of compounds called curcuminoids, which include curcumin (diferuloylmethane), demethoxycurcumin, and bisdemethoxycurcumin. The best studied compound is curcumin, which constitutes 3.14 % (on average) of powdered turmeric. In addition, there are other important volatile oils such as turmerone, atlantone, and zingiberene. Some general constituents are sugars, proteins, and resins. Curcumin, the active component of turmeric, has also been shown to be a vitamin D receptor ligand "with implications for colon cancer chemoprevention" (Fig. 4.1ag) [421–423].

Fig. 4.1ah *Buddleja globosa*

Buddleja globosa, also known as the orange ball buddleja, is a species of flowering plants endemic to Chile and Argentina, where it grows in dry and moist forest. The young branches are subquadrangular and tomentose, bearing sessile or subsessile lanceolae or elliptic leaves that are 5–15 cm long by 2–6 cm wide, glabrescent and bullate above and tomentose below. Flowers are yellow to orange, 1.2–2.8 cm in diameter, and heavily honey scented. Folk medicine attributes to *B. globosa* wound-healing properties, and the infusion of the leaves is used topically for the treatment of wounds, burns, and external and internal ulcers. Chemical studies of this species have allowed to isolate glycosidic flavonoids, phenylethanoids including verbascoside, iridoids, triterpenoids, diterpenoids, and sesquiterpenoids (Fig. 4.1ah) [424–426].

Wolfiporia extensa (Peck) Ginns is a fungus in the Polyporaceae family. It is a wood-decay fungus but has a terrestrial growth habit. It is notable in the development of a large, long-lasting underground sclerotium that resembles a small coconut. This sclerotium called tuckahoe, or Indian bread, was used by Native Americans as a source of food in times of scarcity. It is also used as a medicinal mushroom in Chinese medicine (Fig. 4.1ai).

Helianthus annuus (sunflower) is a species of flowering plants belonging to Asteraceae family. It is native to North America. The genus is one of many in the Asteraceae that are known as sunflowers. *Helianthus* species are used as food plants by the larvae of many Lepidoptera species. They bear one or several wide, terminal capitula with bright yellow ray florets at the outside and yellow or maroon disk florets inside. They usually grow to the height of 50–390 cm or more. Stems are rough and hairy; leaves are dentate and often sticky. The domesticated sunflower, *Helianthus annuus*, is the most familiar species (Fig. 4.1aj).

Tanacetum parthenium is a traditional medicinal herb also known as feverfew which is commonly used to prevent migraine headaches and is also occasionally

Fig. 4.1ai *Wolfiporia extensa*

Fig. 4.1aj *Helianthus annuus*

grown for ornament. The plant grows into a small bush up to around 46 cm high with citrus-scented leaves and is covered by flowers reminiscent of daisies. It spreads rapidly, and they will cover a wide area after a few years. Feverfew was native to Eurasia, specifically the Balkan Peninsula, Anatolia, and the Caucasus, and also found in the rest of Europe, North America, and Chile. Feverfew has been used as a herbal treatment to reduce fever and to treat headaches, arthritis, and digestive problems, though scientific evidence does not support anything beyond a

Fig. 4.1ak *Tanacetum parthenium*

placebo effect. The active ingredients in feverfew include parthenolide and tanetin. There has been some scientific interest in parthenolide, which has been shown to induce apoptosis in some cancer cell lines in vitro and potentially to target cancer stem cells (Fig. 4.1ak) [427–429].

Ribes nigrum is a woody shrub in the family Grossulariaceae grown for its piquant berries. It is native to temperate parts of central and northern Europe and northern Asia. *Ribes nigrum*, the blackcurrant, is a medium-sized shrub, growing to 1.5 m. The leaves are alternate, simple, 3–5 cm broad, and long with five palmate lobes and a serrated margin. Blackcurrant fruit is rich in vitamin C, various other nutrients, phytochemicals and antioxidants, and seed oil is also rich in many nutrients, especially vitamin E and several unsaturated fatty acids including alpha-linolenic acid and gamma-linolenic acid. Major anthocyanins in blackcurrant pomace are delphinidin-3-O-glucoside, delphinidin-3-O-rutinoside, cyanidin-3-O-glucoside, and cyanidin-3-O-rutinoside which are retained in the juice concentrate among other yet unidentified polyphenols. It has potential to inhibit inflammation, cancer, microbial infections, or neurological disorders like Alzheimer's disease, and blackcurrant juice was found to have high antioxidant content and a potent free radical scavenger. The fruit is also used in the preparation of alcoholic beverages, and both fruit and foliage have uses in traditional medicine and the preparation of dyes. In Europe, the leaves have traditionally been used for arthritis, spasmodic cough, and diarrhea and as a diuretic and for treating a sore throat. The berries were made into a drink thought to be beneficial for treatment of colds and flu, for other fevers, for diaphoresis, and as a diuretic (Fig. 4.1al) [430–434].

Citrus sinensis is the fruit of the citrus species in the family Rutaceae. The fruit of the *Citrus sinensis* is considered a sweet orange. It was probably originating in Southeast Asia, cultivated in China from ancient times. Orange trees are widely grown in tropical and subtropical climates for their sweet fruit. The fruit of the

Fig. 4.1al   *Ribes nigrum*

Fig. 4.1am   *Citrus sinensis*

orange tree can be eaten fresh or processed for its juice or fragrant peel. Sweet oranges currently account for approximately 70 % of citrus production (Fig. 4.1am).

Atractylodes lancea, a member of the Compositae family, is a traditional Chinese medicinal plant. Volatile oils from *A. lancea* show antimicrobial activities as well. These oils comprise active secondary metabolites terpenes, flavonoids, and alkaloids including the characteristic components atractylone, β-eudesmol, hinesol, and atractylodin. Activation of multiple signaling events by plant as defense response, i.e., jasmonic acid biosynthesis by plants, is induced by pathogen infection and salicylic acid is involved in activating distinct sets of defense-related genes (Fig. 4.1an) [435–438].

Fig. 4.1an   *Atractylodes lancea*

Fig. 4.1ao   *Idesia polycarpa*

Idesia polycarpa is a species of flowering plants in the family Salicaceae (formerly Flacourtiaceae). It is native to eastern Asia in China, Japan, Korea, and Taiwan. It is a medium-sized deciduous tree reaching a height of 8–21 m, with a trunk up to 50 cm diameter with smooth grayish-green bark. The leaves are large, heart shaped, 8–20 cm long, and 7–20 cm broad, with a red 4–30 cm petiole bearing two or more glands. The flowers are small, yellowish green, fragrant, and 13–30 cm long. The fruit is a berry 5–10 mm diameter, ripening orange to dark purple red, containing several 2–3 mm brown seeds, and often persisting until the following spring (Fig. 4.1ao).

Fig. 4.1ap   *Ipomoea batatas*

Ipomoea batatas (sweet potato) is a dicotyledonous plant that belongs to the family Convolvulaceae. Roots are large, tuberous, starchy, and sweet tasting. The young leaves and shoots are sometimes eaten as greens. *I. batatas* is the only crop plant of major importance; some others are used locally, but many are actually poisonous. The plant is an herbaceous perennial vine, bearing alternate heart-shaped or palmately lobed leaves and medium-sized sympetalous flowers. The edible tuberous root is long and tapered, with a smooth skin whose color ranges between yellow, orange, red, brown, purple, and beige (Fig. 4.1ap) [439].

Atractylodes macrocephala is one of the oldest and most frequently used Chinese herbs for oriental medicine in China. Many components, such as volatile oils, sesquiterpenoids, polysaccharides, amino acids, vitamins, resins and other ingredients, have been found in *Atractylodes macrocephala* up till recently. *A. macrocephala* possesses various bioactivities, including diarrhea, abdominal pain, and insufficiency of the stomach, intestine, liver, and kidney or insufficiency of the spleen with abundance of dampness (Fig. 4.1aq) [440].

Ganoderma is a polypore mushrooms which grows on wood, and nearly 80 species have been reported. It belongs to the family Ganodermataceae. It is an important genus economically based on its traditional medicinal use in Asian Countries. *G. colossum* is popularly known as shelf mushrooms or bracket fungi. *Ganoderma* are characterized by basidiocarps that are large, perennial, woody brackets also called "conks." They are lignicolous and leathery either with or without a stem. The fruit bodies typically grow in a fan-like or hoof-like form on the trunks of living or dead trees. They have double-walled, truncate spores with yellow to brown ornamented inner layers. Ganoderma are wood-decaying fungi with a cosmopolitan distribution. They can grow on both coniferous and hardwood species. They are white-rot fungi with enzymes that allow them to break down wood components such as lignin and cellulose. Significant research has been trying to harness the

Fig. 4.1aq   *Atractylodes macrocephala*

Fig. 4.1ar   *Ganoderma colossum*

power of these wood-degrading enzymes for industrial applications such as biopulp-
ing or bioremediation (Fig. 4.1ar) [441].

Ganoderma lucidum is a member of Ganodermataceae family. *Ganoderma* is a
genus of polypore mushrooms which grow on wood. It contains bioactive compounds such as triterpenoids and polysaccharides useful in treatment of various
diseases. Moreover, *G. lucidum* contains the largest variety of cellulose, lignin, and
xylan-digesting enzymes, which are being used in biomass remediation and industrial sludge processing. *G. lucidum* contains variety of potential therapeutic benefits

like anticancer, immunoregulatory, antioxidant, hepatoprotective, hypoglycemic, antibacterial, antiviral, antifungal, and reducing blood cholesterol (Fig. 4.1as) [442, 443].

Justicia is a genus of flowering plants in the family Acanthaceae. The plant is native to warm temperate regions of the America, with two species occurring north into cooler temperate regions. Common names include water-willow and shrimp plant, the latter from the inflorescences, which resemble a shrimp in some species. They are evergreen perennials and shrubs with leaves which are often strongly veined; but they are primarily cultivated for their showy tubular flowers in shades of white, cream, yellow, orange or pink (Fig. 4.1at).

Fig. 4.1as *Ganoderma lucidum*

Fig. 4.1at *Justicia ciliate*

Fig. 4.1au *Vaccinium macrocarpon*

Vaccinium macrocarpon is also called large cranberry, American cranberry, and bearberry. It is a cranberry of the subgenus *Oxycoccus* and genus *Vaccinium*. It is native to North America. Preliminary studies show *V. macrocarpon* fruit has antibacterial activity against the intestinal pathogens *Escherichia coli* and *Listeria monocytogenes* (Fig. 4.1au).

Brosimum acutifolium belongs to the family Moraceae. It is also known as ahua jonra, mururi, takini, tamamuri, tauni, and vegetable mercury. It has been used for arthritis and rheumatism. Researchers found that it is also helpful in reducing inflammation. The active compound mururin A and B have the ability to inhibit protein kinase C (PKC) and protein kinase A (PKA). PKC is involved with various conditions and is one of the chemicals that the body uses to actually produce inflammation. People with autoimmune disorders, arthritis, and rheumatoid arthritis usually have elevated PKC levels, and PKC inhibitors are a new class of drugs under research for these types of conditions. PKC and PKA also play role in cancer and tumor cell growth. Researchers at Cornell University reported that tamamuri bark showed in vitro antibacterial actions against *Bacillus* and *Staphylococcus* (Fig. 4.1av) [444].

Ruscus aculeatus (Butcher's broom) is a low evergreen Eurasian shrub, with flat shoots known as cladodes that give the appearance of stiff, spine-tipped leaves. Small greenish flowers appear in spring, and are borne singly in the center of the cladodes. *Ruscus aculeatus* occurs in woodlands and hedgerows, where it is tolerant of deep shade, and also on coastal cliffs. It is also widely planted in gardens and has spread as a garden escape in many areas outside its native range. *R. aculeatus* has been known to enhance blood flow to the brain, legs, and hands. It has been used to relieve constipation and water retention and improve circulation. It is also used to treat varicose veins and treat hemorrhoids. Suggested mechanisms to explain this include stimulation of venous alpha 1 and 2 adrenoreceptors and decreased capillary permeability (Fig. 4.1aw) [445, 446].

Fig. 4.1av *Brosimum acutifolium*

Fig. 4.1aw *Ruscus aculeatus*

Morus nigra, the black mulberry, is a species of flowering plants in the family Moraceae, native to southwestern Asia. It is known for its large number of chromosomes, as it has 154 pairs. Morus nigra is a deciduous tree growing to 12 m tall by 15 m broad. The leaves are 10–20 cm long by 6–10 cm broad—up to 23 cm long on vigorous shoots, downy on the underside, the upper surface rough with very short, stiff hairs. The edible fruit is dark purple, almost black, when ripe, 2–3 cm long, and a compound cluster of several small drupes; it is richly flavored, similar to the red mulberry (*Morus rubra*) but unlike the more insipid fruit of the white mulberry (*Morus alba*) (Fig. 4.1ax).

Fig. 4.1ax *Morus nigra*

Fig. 4.1ay *Ornithogalum caudatum*

Ornithogalum caudatum or false sea onion or pregnant onion is a curious bulbous plant and looks very similar to an onion. It cannot tolerate the dry period of the Mediterranean climate. The bulb may reach up to 10 cm in diameter. Leaves under ideal circumstances grow up to 40 cm in length. They are wonderful novelty plants for indoors or out. It looks like a pregnant mother; the bulb grows on the top of the soil. The babies grow under the skin making the bulb look as if it is pregnant. Once the little bulblets free themselves, they sit on top of the soil and they will root and grow. Traditionally, it has been tide over cuts and bruises and said to have healing effects similar to aloe vera (Fig. 4.1ay).

Chenopodium quinoa is an annual herbaceous plant in the family Amaranthaceae. It was discovered as a health food by North Americans and Europeans and has increased dramatically in popularity in recent years because it is gluten-free and high in protein. It is a grain crop grown primarily for its edible seeds. Quinoa is a fast-growing plant, up to 2 m tall, with alternate and coarsely toothed leaves. Quinoa can be cooked to produce a fluffy grain-like dish with a nut-like flavor, although the seed coats contain saponins that convey a bitter flavor unless the achenes are rinsed and soaked before use. Quinoa was traditionally fermented into a beer-like beverage, chichi, by the Incas. The leaves can be used as a cooked vegetable, similar to spinach. It is similar to *Chenopodium* species, such as pitseed goosefoot (*Chenopodium berlandieri*) and fat hen (*Chenopodium album*), which were grown and domesticated in North America as part of the Eastern Agricultural Complex before maize agriculture became popular. The nutrient composition is favorable compared with common cereals. Quinoa seeds contain essential amino acids like lysine and acceptable quantities of calcium, phosphorus, and iron (Fig. 4.1az).

Camellia sinensis is a type of tea called black tea. It is more oxidized and stronger in flavor than other types of tea. Black tea contains negligible quantities of calories, protein, sodium, and fat. Camellia tea plants are rich in polyphenols, which are a type of antioxidant. Drinking a moderate amount of black tea may boost blood pressure slightly, but the effect does not last long (Fig. 4.1aaa) [447].

Prunus amygdalus is a species of tree native to the Middle East and South Asia. It is also known as almond. The fruit of the almond is a drupe, consisting of an outer hull and a hard shell with the seed inside. Shelling almonds refers to removing the shell to reveal the seed. The almond is a deciduous tree, growing 4–10 m in height, with a trunk of up to 30 cm in diameter. The leaves are 3–5 in. long, with a serrated margin and a 2.5 cm petiole. The flowers are white to pale pink, 3–5 cm diameter

Fig. 4.1az *Chenopodium quinoa*

Fig. 4.1aaa *Camellia sinensis*

Fig. 4.1aab *Prunus amygdalus*

with five petals, produced singly or in pairs, and appearing before the leaves in early spring (Fig. 4.1aab) [448].

Andrographis paniculata is an annual herbaceous plant in the family Acanthaceae, native to India and Sri Lanka. It is widely cultivated in Southern and Southeastern Asia, where it has been traditionally used to treat infections and some diseases. Andrographis paniculata grows erect to a height of 30–110 cm in moist, shady places. The lance-shaped leaves have hairless blades measuring up to 8 cm long by 2.5 wide. The small flowers are borne in spreading racemes. The fruit is a capsule

Fig. 4.1aac *Andrographis paniculata*

around 2 cm long and a few millimeters wide. It contains many yellow-brown seeds. The herb has a number of purported medicinal uses, although research has found evidence of its effectiveness is limited to treatment of upper respiratory infection, ulcerative colitis, and rheumatic symptoms; in particular, there is no evidence of its effectiveness in cancer treatment (Fig. 4.1aac) [409].

Pulsatilla cernua is a species of herbaceous perennials native to meadows and prairies of North America, Europe, and Asia. Common names include pasque flower, wind flower, prairie crocus, Easter flower, and meadow anemone. Several species are valued ornamentals because of their finely dissected leaves, solitary bell-shaped flowers, and plumed seed heads. *Pulsatilla* is highly toxic and produces cardiogenic toxins and oxytoxins which slow the heart in humans. Excess use can lead to diarrhea, vomiting and convulsions, hypotension, and coma. Blackfoot Indians used it to induce abortions and childbirth but not be taken during pregnancy nor during lactation. Extracts of *Pulsatilla* have been used to treat reproductive problems such as premenstrual syndrome and epididymitis. Additional applications of plant extracts include their uses as sedatives and treatment of coughs (Fig. 4.1aad) [410].

Tinospora cordifolia, which is known by the common name Guduchi, is an herbaceous vine of the family Menispermaceae, indigenous to the tropical areas of India, Myanmar, and Sri Lanka. The plant is a glabrous climbing shrub found throughout India, typically growing in deciduous and dry forests. The leaves are heart shaped. The succulent bark is creamy white to gray in color, with deep clefts spotted with lenticels. It puts out long, slender aerial roots and is often grown on mango or neem trees. Flowers are yellow, growing in lax racemes from nodes on old wood. Fruits are drupes, turning red when ripe. The active adaptogenic constituents are diterpene compounds, polyphenols, and polysaccharides, including arabinogalactan polysaccharide. A research concluded that *T. cordifolia* exhibited hepatoprotective property (Fig. 4.1aae) [411, 449].

Fig. 4.1aad   *Pulsatilla cernua*

Fig. 4.1aae   *Tinospora cordifolia*

Rheum emodi is a species of perennial plants in the family Polygonaceae. The species have large somewhat triangular-shaped leaves with long, fleshy petioles. The flowers are small, greenish-white to rose-red, and grouped in large compound leafy inflorescences. Many *Rheum* species have food and medicinal uses. *Rheum rhabarbarum* is used to make pies, jellies, jams, and wine. All parts of the plant contain the poison oxalic acid, but its concentration in the leaf stems or petioles used in food preparation is very low, and their tart flavor instead is caused by nontoxic malic acid. The plants also produce other poisonous compounds, including citric acid and anthraquinone glycosides, and the raw or cooked leaf blades are poisonous to humans and livestock if consumed in large enough amounts (Fig. 4.1aaf).

Physalis peruviana is also known as Cape gooseberry, Inca berry, Aztec berry, golden berry, giant ground cherry, Peruvian groundcherry, etc. The fruit is a smooth berry, resembling a miniature, spherical, yellow tomato. Removed from its bladder-like calyx, it is about the size of a marble, about 1–2 cm in diameter. Like a tomato, it contains numerous small seeds. It is bright yellow to orange in color, and it is sweet when ripe, with a characteristic, mildly tart flavor, making it ideal for snacks, pies, or jams. It is relished in salads and fruit salads, sometimes combined with avocado. Also, because of the fruit's decorative appearance, it is popular in restaurants as an exotic garnish for desserts (Fig. 4.1aag) [450].

Fig. 4.1aaf   *Rheum emodi*

Fig. 4.1aag   *Physalis peruviana*

Fig. 4.1aah   *Echinochloa utilis*

Echinochloa utilis is a grass genus, some of whose members are millets grown as cereal or fodder crops. Collectively, the members of this genus are called barnyard grasses. In particular, common barnyard grass (*E. crus-galli*) is notorious as a weed. Among the plant pathogens that affect this genus are the sac fungus *Cochliobolus sativus*, which has been noted on common barnyard grass, and rice hoja blanca virus. Both affect many other grass species, in particular most important cereals, and Echinochloa weeds may serve as a reservoir. The fungi *Drechslera monoceras* and *Exserohilum monoceras* have been evaluated with some success as potential biocontrol agents of common barnyard grass in rice fields (Fig. 4.1aah) [451].

Withania somnifera is also known as ashwagandha, Indian ginseng, poison gooseberry, or winter cherry; it is a plant of Solanaceae family. Several other species in the genus *Withania* are morphologically similar. Roots are used for medicinal purposes. Berries and leaves are applied externally to tumors, tubercular glands, carbuncles, and ulcers. The roots are used to prepare the herbal remedy ashwagandha, which has been traditionally used to treat various symptoms and conditions. Ashwagandha exhibits greater clinical benefit in concentration, fatigue, social functioning, vitality, and overall quality of life. It has been useful in stress releasing as well as restoring and increasing energy levels. Studies reported a significant improvement in both cardiovascular and respiratory endurance (Fig. 4.1aai).

Melaleuca alternifolia, commonly known as narrow-leaved paperbark, narrow-leaved tea tree, narrow-leaved ti tree, or snow-in-summer, is a species of tree or tall shrub native to Australia. It grows along streams and on swampy flats and is often the dominant species where it occurs. Leaves are linear, 10–35 mm long, and 1 mm wide. White flowers occur in spikes 3–5 cm long. Small woody, cup-shaped fruit are 2–3 mm in diameter. The indigenous people of eastern Australia used leaves to treat coughs and colds. They also sprinkle leaves on wounds, after which a poultice is applied (Fig. 4.1aaj) [452, 453].

Fig. 4.1aai *Withania somnifera*

Fig. 4.1aaj *Melaleuca alternifolia*

Cymbopogon citratus is commonly known as lemon grass or oil grass; it is a tropical plant from Southeast Asia. In the folk medicine of Brazil, it is believed to have anxiolytic, hypnotic, and anticonvulsant properties. Laboratory studies have shown cytoprotective, antioxidant, and anti-inflammatory properties in vitro, as well as antifungal properties. Lemon grass contains 65–85 % of citral and active ingredients such as myrcene, an antibacterial and pain reliever; citronella; citronellol; and geranilol. Its essence is used for the production of skin care products such as lotions, creams, and facial cleansing toner in its pure form (Fig. 4.1aak) [454].

Fig. 4.1aak *Cymbopogon citrates*

Plantago major is a species of *Plantago*, family Plantaginaceae, and is also known as broadleaf plantain or greater plantain. The plant is native to most of Europe and northern and central Asia. *Plantago major* is one of the most abundant and widely distributed medicinal crops in the world. *Plantago major* is a herbaceous perennial plant with a rosette of leaves 15–30 cm in diameter. Each leaf is oval shaped, 5–20 cm long, and 4–9 cm broad. Flowers are small, greenish brown with purple stamens, produced in a dense spike 5–15 cm long on top of a stem 13–15 cm tall. The active chemical constituents are aucubin, allantoin, ursolic acid, flavonoids, and asperuloside. Broadleaf plantain is also a highly nutritious wild edible that is high in calcium and vitamins A, C, and K. A poultice of the leaves can be applied to wounds, stings, and sores in order to facilitate healing and prevent infection. Plantain has astringent properties, and a tea made from the leaves can be ingested to treat diarrhea and soothe raw internal membranes. Scientific studies have shown that plantain extract has a wide range of biological effects, including wound-healing activity, anti-inflammatory, analgesic, antioxidant, weak antibiotic, immunomodulating, and antiulcerogenic activity. When ingested, the aucubin in plantain leaves leads to increased uric acid excretion from the kidneys and may be useful in treating gout (Fig. 4.1aal) [455].

Phyllanthus emblica (or *Emblica officinalis*) is a deciduous tree of the family Phyllanthaceae. It is also popular with its other names, Indian gooseberry, Dhatrik, or amla. Fruits of *E. officinalis* are reputed to contain high amounts of ascorbic acid, the specific contents are disputed, and the overall antioxidant strength of amla may derive instead from its high density of ellagitannins such as emblicanin A (37 %), emblicanin B (33 %), punigluconin (12 %), and pedunculagin (14 %). It also contains punicafolin and phyllanemblinin A, phyllanemblin and other polyphenols: flavonoids, kaempferol, ellagic acid, and gallic acid.

Fig. 4.1aal *Plantago major*

Indian gooseberry has reported activities of antiviral, antimicrobial, anticancer, etc. It also promoted the spontaneous repair and regeneration process of the pancreas occurring after an acute attack. Experimental preparations of leaves, bark, or fruit have shown potential efficacy against laboratory models of disease, such as for inflammation, cancer, age-related renal disease, and diabetes. A study demonstrated that it reduces blood cholesterol levels in both normal and hypercholesterolemic men along with reduction of blood glucose level. In traditional Indian medicine, dried and fresh fruits of the plant are used. According to Ayurveda, amla fruit is sour (amla) and astringent (kashaya) in taste (rasa), with sweet (madhura), bitter (tikta), and pungent (katu) secondary tastes (anurasas). Its qualities (gunas) are light (laghu) and dry (ruksha), the postdigestive effect (vipaka) is sweet (madhura), and its energy (virya) is cooling (shita). *Emblica officinalis* tea may ameliorate diabetic neuropathy due to aldose reductase inhibition. In rats, it significantly reduced blood glucose, food intake, water intake, and urine output in diabetic rats compared with the nondiabetic control group (Fig. 4.1aam) [456–458].

Carpesium longifolium is a species of flowering plants belonging to the family Asteraceae. They are distributed in Europe and Asia. These are mainly perennial herbs, but a few species are annuals. The fruit is a hairless, ribbed, beaked achene. Several other species including *C. abrotanoides*, *C. divaricatum*, and *C. rosulatum* have been used in traditional medicine in China and Korea (Fig. 4.1aan).

Paeonia suffruticosa, the tree peony, is a species of peony native to China. More commonly, the plant is referred to as the tree peony. It is known as mudan in Chinese and is an important symbol in Chinese culture. The flower component is its most attracting feature of the plant. *Paeonia suffruticosa's* flower is very large in comparison with most other flower species. The size of the flower usually ranges from 6 to 12 in. It is the flower component that is used to classify the plant's cultivar, using

Fig. 4.1aam *Phyllanthus emblica*

Fig. 4.1aan *Carpesium longifolium*

characteristics such as the flower form and color. The epidermis of its roots is used in traditional Chinese medicine, called mudanpi (Fig. 4.1aao) [459].

Panax ginseng is any one of 11 species of slow-growing perennial plants with fleshy roots, belonging to the genus *Panax* of the family Araliaceae. Ginseng is found only in the Northern Hemisphere, in North America, and in eastern Asia. Ginseng is characterized by the presence of ginsenosides. Folk medicine attributes various benefits to oral use of *P. ginseng* roots, including roles as an aphrodisiac,

Fig. 4.1aao *Paeonia suffruticosa*

stimulant, type II diabetes treatment or cure for sexual dysfunction in men. The common adaptogen property of *P. ginseng* is generally considered to be relatively safe even in large amounts.

A common side effect of *P. ginseng* may be insomnia, but this effect is disputed. Other side effects can include nausea, diarrhea, headaches, nose bleeds, high blood pressure, low blood pressure, and breast pain. Ginseng may also lead to induction of mania in depressed patients who mix it with antidepressants. Ginseng has been shown to have adverse drug reactions with phenelzine and warfarin; it has been shown to decrease blood alcohol levels. One of the most common and characteristic symptoms of acute overdose of *Panax ginseng* is bleeding. Symptoms of mild overdose may include dry mouth and lips, excitation, fidgeting, irritability, tremor, palpitations, blurred vision, headache, insomnia, increased body temperature, increased blood pressure, edema, decreased appetite, dizziness, itching, eczema, early morning diarrhea, bleeding, and fatigue (Fig. 4.1aap) [460, 461].

Caulophyllum thalictroides is a flowering plant in the Berberidaceae (barberry) family, also called squaw root or papoose root. It is found in hardwood forest of the eastern United States and favors moist coves and hillsides, generally in shady locations, in rich soil. It grows in eastern North America, from Manitoba and Oklahoma east to the Atlantic Ocean. It is a medium-tall perennial with blue berry-like fruits and bluish-green foliage. From the single stalk rising from the ground, there is a single, large, three-branched leaf plus a fruiting stalk. The bluish-green leaflets are tulip shaped, entire at the base, but serrate at the tip. It is used as a medicinal herb by American Indians. Many Native American tribes use this herb in conjunction with other herbs and fluids for abortive and contraceptive purposes (Fig. 4.1aaq) [462].

Fig. 4.1aap   *Panax ginseng*

Fig. 4.1aaq   *Caulophyllum thalictroides*

Bidens pilosa is a species of flowering plants in the family Asteraceae. It is a native plant of South America and has cosmopolitan distribution; in some parts of the world, it is a source of food. It is considered a weed in some tropical habitats. However, in some parts of the world, it is a source of food. Due to the type of inflorescence present they are called chapter disk flowers; flowers are fertile and yellow. The seeds are black and stick to livestock (Fig. 4.1aar).

Fig. 4.1aar *Bidens pilosa*

4.2 Marine Sources

Marine resources are one of the richest sources of anti-inflammatory medicines; these are biological entities found in seas and oceans that are beneficial to man. Marine sources contain a huge biodiversity including fish, coral reefs, crabs, microorganisms, and fungi of several species with significant medicinal importance. An effort is required to conserve and protect these resources from destruction of extinction, mainly because of human activities like pollution and over fishing. Table 4.2 contains a vast anti-inflammatory source found in sea and oceans.

Aspergillus terreus is a fungus found worldwide in soil. It is also known as *Aspergillus terrestris*. This saprophytic fungus is prevalent in warmer climates such as tropical and subtropical regions. Aside from being located in soil, *A. terreus* has also been found in habitats such as decomposing vegetation and dust. *Aspergillus terreus* produces a number of secondary metabolites and mycotoxins including territrem A, citreoviridin, citrinin, gliotoxin, patulin, terrein, terreic acid, and terretonin. Aspergillus terreus is commonly used in industry to produce important organic acids, such as itaconic acid and cis-aconitic acid as well as enzymes, like xylanase. *Aspergillus terreus* can cause opportunistic infection in people with deficient immune systems. It is relatively resistant to amphotericin B, a common antifungal drug. It causes both systemic and superficial infections. It causes typical respiratory infection. *Aspergillus terreus* has the ability to cause serious effects in immunocompromised patients who lack specific immune cells. Specifically, it is prolonged neutropenia that predisposes humans and animals to this fungal disease. It hinders cholesterol synthesis by inhibiting enzymes responsible for the process (Fig. 4.2a) [548].

Table 4.2 Anti-inflammatory marine source

Organism	Source	Compound	Activity	Reference
Acanthostrongylophora	Sponge	12,34-Oxamanzamine E	Neuritogenic activity	[463, 464]
Ancorina sp.	Sponge	Ancorinolates A–C and bis-ancorinolate B	Weak HIV-inhibitory activity	[465]
Aplidium aff. densum	Tunicate	Methoxyconidiol and didehydroconicol	Biological activities against bacteria and human lymphoblastic cell lines	[466]
Aspergillus terreus	Ascomycoya	Terretonins A–D and asterrelenin	Antioxidant activity	[467, 468]
Auxarthron reticulatum	Fungus	Amauromine and penicinoline	Cytotoxic and antimicrobial activities	[469]
Axinella carteri	Marine sponge	Hymenialdisine	Cytokine inhibitory activity	[470, 471]
Briareum excavatum	Gorgonian	Excavatolides A–E and U–Z and briaexcavatolides S–V	Anti-HCMV (human cytomegalovirus) cytotoxic activity	[472–476]
Briareum stechei	Octocoral	Milolides, solenolide C	Analgesic activity	[477]
Bursatella leachii	Sea hare	Malyngamide	Cytotoxicity activity	[478]
Callipeltas sp.	Sponge	Callipeltin A	Cytotoxic activity	[479]
Ceratodictyon spongiosum	Marine algae	Ceratospongamide	Cytotoxic activity	[480, 481]
Dendronephthya sp.	Octocoral	Isogosterones A–D	Antibacterial activity	[482, 483]
Dictyochloris fragrans	Marine alga	Sulfonoquinovosyl dipalmitoyl glyceride	Antiviral activity	[484]
Dragmacidon sp.	Sponge	Dragmacidonamines A and B	Cytotoxic, antitumor activity	[485]
Eleutherobia aurea	Soft coral	Zahavin	Antiangiogenic activity	[486, 487]
Eunicea sp.	Octocoral	Calyculaglycosides D and E, (±)-nephthenol	Antiplasmodial activity	[488, 489]
Fasciospongia cavernosa	Sponge	Cacospongionolide E	Antimicrobial activity	[490–492]
Glossodoris atromarginata	Dorid nudibranch	Scalaranes	Cytotoxic activity	[493]
Glossodoris nudibranchs	Sponge	Homoscalarane, scalarane compounds	Cytotoxic activity	[494, 495]
Haematococcus pluvialis	Microalgae	Astaxanthin	Antioxidant activity	[496]
Hyrtios erectus	Marine sponge	Homofascaplysin A and fascaplysin, salmahyrtisol A, B, and sesterstatin 16-hydroxyscalarolide	Antiproliferative activity, cytotoxic activity, antifungal activity, antiplasmodial activity	[497–501]

Ianthella basta	Hemibastadins, hemibastadinols	Sponge	Cytotoxic, antifouling, antitumoral, antibiotic, antiviral, enzyme-inhibitory activity	[502–505]
Ianthella quadrangulata	Iso-iantheran A	Sponge	Cytotoxic activity	[506, 507]
Junceella fragilis	Deacetyljunceellolide D	Coral	Mitochondrial activity	[508, 509]
Junceella juncea	Junceelolide B, C, and D	Coral	Antifeedant and antifouling activity	[510, 511]
Lobophytum crassum	Cembrane (2S,7S,8S)-sarcophytoxide	Soft coral	Antimicrobial and HIV-inhibitory activity	[512, 513]
Nostoc commune	Dodecahydrophenanthrene	Terrestrial cyanobacterium	Antibacterial activity	[514]
Pachyclavularia violacea	Pachyclavulides A, B, C, and D	Soft coral	Cytotoxic activity	[515]
Phyllospongia madagascariensis	Norscalarane	Madagascan sponge	Cytotoxic activity	[516]
Poria cocos	Pachymic acid, dehydrotumulosic acid	Fungus	Nematicidal activity	[517, 518]
Pseudopterogorgia bipinnata	Kallolide A, bipinnatin A, C, J, and bipinnapterolide A	Octacoral	Antiplasmodial Activity	[519, 520]
Pseudopterogorgia elisabethae	Elisabethins A–C, D and elisapterosin A and B amphilectane-type diterpenes	Gorgonian octocoral	Cytotoxic and analgesic activity antituberculosis activity	[521–524]
Pseudopterogorgia kallos	Bipinnatins E, K–Q	Octacoral	Analgesic activity	[525]
Renilla reniformis	Renillins A–D	Sea pansy	Luciferase activity	[526, 527]
Rhaphisia lacazei	Bisindole	Sponge	Antiproliferative activity	[528]
Sarcophyton crassocaule	Sarcrassins A–E	Soft coral	Cytotoxic activity	[529, 530]
Sarcophyton ehrenbergi	Sarcophytol T	Soft coral	Antiviral activity	[531, 532]
Serratia ureilytica	Serlyticin-A	Bacteria	Antioxidant activity	[533]
Streptomyces species	Cyclomarins	Marine bacterium	Antibacterial activity	[534–536]
Stylissa caribica	Stylisin	Sponge	Antimicrobial, antimalarial, anticancer, anti-HIV-1, anti-Mtb, activity	[537]

(continued)

Table 4.2 (continued)

Stylissa flabellata	Sponge	Stylissadines A and B	Cytotoxic activity	[538]
Tolypothrix nodosa	Cyanobacterium	Tolypodiol	Anticancer	[539, 540]
Tricleocarpa fragilis	Algae	Cycloartenol sulfates and nor-lanosterol sulfates	–	[541]
Vibrio parahaemolyticus	North sea bacterium	Arundine	Hemolytic activity	[542, 543]
Xylaria sp.	Fungus	Kolokosides A–D	Antimicrobial activity	[544, 545]
Zostera japonica	Sea grass	Palmitic acid	Photosynthetic activity	[546, 547]

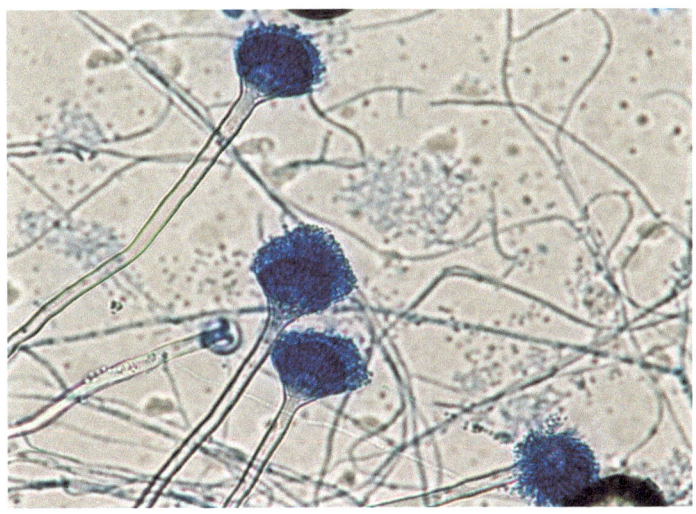

Fig. 4.2a *Aspergillus terreus*

Fig. 4.2b *Briareum excavatum*

Briareum excavatum is a gorgonian coral belonging to the genus *Briareum* (*Briareidae*) which were proven to be a rich source of marine diterpenoid derivatives, such as briarane, asbestinane, and eunicellin-type natural products. Most compounds of these classes were found to possess complex structure. Excavatoids A–D, a new polyoxygenated briaranes, are obtained from the octocoral *Briareum excavatum* (Fig. 4.2b) [549].

Ceratodictyon spongiosum is a woolly branching organism. Grows on coral rubble, it looks like a sponge, complete with holes along the stems. It is overall 20–30 cm in length with stem 1–1.5 cm wide. It is light to dark green in color. The

Fig. 4.2c   *Ceratodictyon spongiosum*

stems are often dividing frequently to form large spreading mats with tips usually y shaped; each stem is made up of fine, branched filaments that are packed together to form structures that feel woolly, velvety, spongy, or felt like with tiny holes on each stems. This organism is actually a symbiotic combination of algae (*Ceratodictyon spongiosum*) and a sponge (*Haliclona cymaeformis*). The algae make up the bulk of the organism, while the sponge provides the tough exterior and appears to give the organism its shape and form, contributing to the formation of the tiny holes. The algae get most of the nitrogen it needs from the sponge (Fig. 4.2c) [550].

Dendronephthya species collectively reaches up to 250 in number. It is found in a variety of flamboyant colors with red or orange being the most common. It is normally shipped while attached to a small piece of live rock or coral rubble. The Carnation Tree Coral or Dendronephthya Carnation is one of the most beautiful and peaceful corals and is also known as the cauliflower soft coral, or strawberry soft coral. The Carnation Tree Coral requires excellent water conditions with very low nitrate and phosphate levels as well as frequent feedings of phytoplankton-based foods. This combination is difficult to achieve for most hobbyists, which makes this species the domain of the expert hobbyist. Unlike many other species, the Carnation Tree Coral does not contain symbiotic algae zooxanthellae; thus, it needs to get all of its nutrition from micro-foods that it catches from the water currents. Ideally, they should be able to filter feed on nutrient-rich waters containing phytoplankton and photosynthesizing microscopic organisms that are found inhabiting the upper sunlight areas of the ocean (Fig. 4.2d).

Doriprismatica atromarginata is a species of sea slug, a dorid nudibranch. It is a shell-less marine gastropod mollusk in the family Chromodorididae. It ranges in color from creamy white through yellow to pale brown. It typically has a black-lined edge running down the outside of a much folded mantle and black rhinophore

Fig. 4.2d *Dendroephthya* sp.

Fig. 4.2e *Doriprismatica atromarginata*

clubs. The frilly mantle sometimes appears to move like a wave as the animal crawls along. It can reach a total length of at least 60 mm (Fig. 4.2e) [551–553].

Haematococcus pluvialis is a freshwater species of *Chlorophyta* from the family Haematococcaceae. This species is well known for its high content of the strong antioxidant astaxanthin, which is important in aquaculture and cosmetics. The high amount of astaxanthin is present in the resting cells, which are produced and rapidly accumulated when the environmental conditions become unfavorable for normal cell growth. Examples of such conditions include bright light, high salinity, and low availability of nutrients. *Haematococcus pluvialis* is usually found in temperate

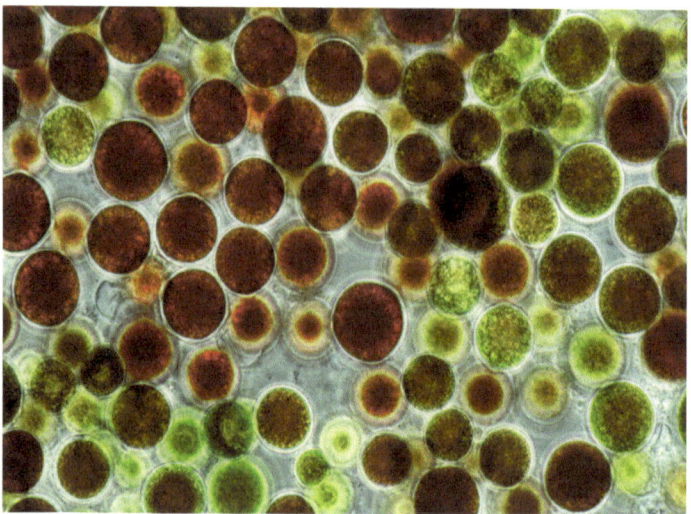

Fig. 4.2f *Haematococcus pluvialis*

regions around the world. Their resting cysts are often responsible for the blood-red color seen in the bottom of dried out rock pools and bird baths. This color is caused by astaxanthin which is believed to protect the resting cysts from the detrimental effect of UV radiation, when exposed to direct sunlight (Fig. 4.2f) [554].

Hyrtios erectus belongs to the genus *Hyrtios*. Previous chemical investigations of different *Hyrtios* sp. and their associated microorganisms revealed the presence of numerous structurally unique natural products including scalarane, acyclic triterpenes, indole alkaloids, and macrolides in addition to steroids. Many of these compounds possess different biological activities. Spongistatins, the most important metabolites of the genus *Hyrtios*, shows powerful anticancer activity, which encouraged an exhaustive recollection of the *H. erecta* sponge for further chemical investigations, which afforded the antineoplastic agents sesterstatins (Fig. 4.2g) [555].

Ianthella basta or elephant ear sponge is a species of fan-shaped sponge in the class Demospongiae. It is also known as the paper sponge or scroll sponge. Sponges are marine invertebrates with a jellylike mesohyl sandwiched between two layers of cells. They are filter feeders maintaining a flow of water through their structure which passes out through large openings called oscula. They have a fragile skeleton of stiff spicules. Research is being undertaken on various metabolites and other biologically active constituents that are synthesized by the sponge (Fig. 4.2h).

Pachyclavularia violacea is synonymous to *Clavularia viridis* which belongs to the family Clavularidae. It is a tube coral Stolonifera which grows with a kind of leathery stolon directly on a solid substrate. Sclerites of the lower layer are branched y shaped, the upper layer of straight and spindle shaped. The colored sclerites the stolon is colored purple. On the stolon grow tubes, each of them a polyp. The polyps are usually light or dark brown in color (Fig. 4.2i).

Fig. 4.2g *Hyrtios erectus*

Fig. 4.2h *Ianthella basta*

Phyllospongia madagascariensis is a type of marine organism and *Phyllospongia* genus of the Phyllospongiinae subfamily. It is a sedentary filter feeder, has a porous body with holes and channels that allow water flow throughout, depends on water flow through its body to obtain food and remove waste, has a jelly-like mesohyl between two thin layers of cells, and contains unspecialized cells that can transform into other cells, can migrate between the main cell layers and the mesohyl, and do not have any developed organ systems. They are motile ingest another organisms or their product to live (Fig. 4.2j).

Fig. 4.2i *Pachyclavularia violacea*

Fig. 4.2j *Phyllospongia madagascariensis*

Poria cocos is a fungus in the Polyporaceae family. It is now known as *Wolfiporia extensa* (Peck). It is a famous home-grown bulk Chinese herb, which has been used medicinally for about 2,000 years. It is a wood-decay fungus but has a terrestrial growth habit. It is notable in the development of a large, long-lasting underground sclerotium that resembles a small coconut. *Poria sclerotium* contains chemical constituents like pachymic acid, pachyman, pachymaran, tumulosic acid, polysaccharide, ergosterol, caprylic acid, undecanoic acid, lauric acid, dodecenoic acid,

Fig. 4.2k *Poria cocos*

palmitic acid, dodecanoate, caprylate, and other elements. It is also used as a medicinal mushroom in Chinese medicine. In traditional medicine, poria mushroom filaments have been used for amnesia, anxiety, restlessness, fatigue, tension, nervousness, dizziness, urination problems, fluid retention, insomnia, an enlarged spleen, stomach problems, diarrhea, tumors, and to control coughing. *Poria* mushroom contains chemicals that might improve kidney function, lower serum cholesterol, reduce inflammation, and suppress immune function. It might also have antitumor and anti-vomiting effects (Fig. 4.2k).

Pseudopterogorgia elisabethae, the Caribbean sea plume, is found in the tropical Caribbean portion of the Atlantic Ocean. This species, as well as other gorgonian octocorals, is found most extensively in this region of the world. *Pseudopterogorgia elisabethae* is usually found at deeper and calmer reef sites, up to depths of about 100 ft, and along reef drop offs. *Pseudopterogorgia elisabethae* is frequently found as a bushy aggregation of feather-like branches, each resembling a plume, around a central axis. The branches are long, with pinnate, distichous branchlets. This tall feathery morphology is more suitable for the deeper waters where the water movement is slower because the gentle currents will not uproot the structure. The pseudopterosins that are produced have been used in medications such as analgesics, and as a nonsteroidal anti-inflammatory drug. The main compound isolated is pseudopterosin A, which is being studied because of the selectivity it exhibits as an analgesic (Fig. 4.2l) [522, 556].

Pseudopterogorgia kallos is a marine organism belonging to the family Gorgoniidae. This organism is a type of aquatic animals with cnidocytes to capture prey and a mesoglea jellylike body. They have cnidocytes, which are specialized cells used for capturing prey; bodies consisting of mesoglea between two layers of epithelium that are mostly one cell thick, swimming medusae and sessile polyps,

Fig. 4.2l *Pseudopterogorgia elisabethae*

Fig. 4.2m *Pseudopterogorgia kallos*

radially symmetrical with mouths surrounded by tentacles that bear cnidocytes; and a single orifice and body cavity used for digestion and respiration. It is a multicellular organism and must ingest another organisms or their product to live (Fig. 4.2m).

Renilla reniformis, the sea pansy, is a colonial cnidarian native to warm continental shelf waters of the New World. It is frequently found washed ashore on northeast Florida. It also can often be found living intertidally completely buried in the sand. Its predator is the striped sea slug *Armina tigrina*. The sea pansy is a collection of polyps with different forms and functions. A single, giant polyp up to two

Fig. 4.2n *Renilla reniformis*

inches in diameter forms the anchoring stem. The pansy-like body bears many small, anemone-like feeding polyps. Cluster of tentacle-less polyps forms an outlet valve that releases water to deflate the colony. Small white dots between the feeding polyps are polyps that act as pumps to expand the deflated colony. The feeding polyps secrete sticky mucus to trap tiny organisms suspended in the water. The colony's rigidity and purple color come from calcium carbonate spicules throughout the polyp's tissues (Fig. 4.2n).

Sarcophyton crassocaule is a type of marine organism belonging to the family Alcyoniina. This organism is a type of aquatic animals with cnidocytes to capture prey and a mesoglea jellylike body. They have cnidocytes, which are specialized cells used for capturing prey; bodies consisting of mesoglea between two layers of epithelium that are mostly one cell thick, swimming medusae and sessile polyps, radially symmetrical with mouths surrounded by tentacles that bear cnidocytes; and a single orifice and body cavity used for digestion and respiration. It is a multicellular organism and must ingest another organisms or their product to live (Fig. 4.2o).

Serratia ureilytica belongs to genus *Serratia* of the family Enterobacteriaceae. It is Gram-negative, facultatively anaerobic, rod-shaped, urea-dissolving, and nonspore-forming bacterium. Members of this genus produce characteristic red pigment, prodigiosin, and can be distinguished from other members of the family Enterobacteriaceae by their unique production of three enzymes: DNase, lipase, and gelatinase. *Serratia* infection is responsible for about 2 % of nosocomial infections of the bloodstream, lower respiratory tract, urinary tract, surgical wounds, and skin and soft tissues in adult patients. *Serratia* infection has caused endocarditis and osteomyelitis in people addicted to heroin (Fig. 4.2p).

Stylissa caribica is a type of marine organism, in the family Dictyonellidae. *Stylissa caribica* is a sponge and strictly marine, is a sedentary filter feeder, has a

Fig. 4.2o   *Sarcophyton crassocaule*

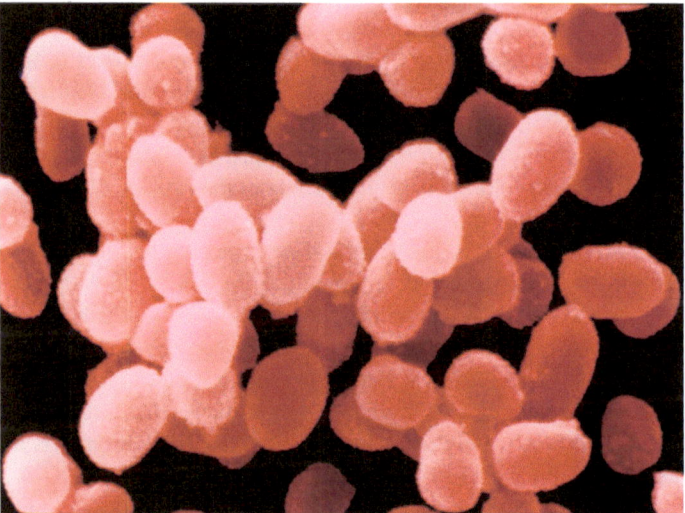

Fig. 4.2p   *Serratia ureilytica*

porous body with holes and channels that allow water flow throughout, depends on water flow through its body to obtain food and remove waste, has a jellylike meso-hyl between two thin layers of cells, and contains unspecialized cells that can transform into other cells, can migrate between the main cell layers and the mesohyl, and do not have any developed organ system (Fig. 4.2q).

Tricleocarpa fragilis is a type of marine organism of the family Galaxauraceae. It is a multicellular thallus with apical growth and has tetrasporangia, filamentous

Fig. 4.2q   *Stylissa caribica*

Fig. 4.2r   *Tricleocarpa fragilis*

gonimoblast, and pit connections. *T. fragilis* cells have multiple nuclei and plastids, triphasic alternation of generations life history. It is red algae with eukaryotic cells without flagella and centrioles using floridean polysaccharides as food reserve. These are multicellular organisms with accessory red pigments known as phycobili-proteins; however chloroplasts lacks external endoplasmic reticulum and containing unstacked thylakoids. Their cell wall is made up of cellulose, they obtain energy from sun, and perform sexual or asexual reproduction with modular and indetermi-nate growth (Fig. 4.2r).

Xylaria is a genus of ascomycetous fungi commonly found growing on dead wood. The name comes from the Greek xýlon meaning wood. Two of the common species of the genus are *Xylaria hypoxylon* and *Xylaria polymorpha. Xylaria hypoxylon* is also known as stag's horn and candle-snuff fungus. It is the most conspicuous because of its erect, 3–7 cm tall, antler-like ascocarps (fruit bodies) which are black at the base but white and branched towards the top, where the fruiting bodies produce white conidia. *Xylaria polymorpha*, commonly known as dead man's fingers, is a saprobic fungus. It often grows in finger-like clusters from the base of a tree or from wood just below ground level. This is a primary fungus utilized in the spalting of sugar maple and other hardwoods. It is a common inhabitant of forest and woodland areas, usually growing from the bases of rotting or injured tree stumps and decaying wood. A variety of bioactive compounds have been identified in this fungus. The compounds xylarial A and B both have moderate cytotoxic activity against the human hepatocellular carcinoma cell line Hep G2. The pyrone derivative compounds named xylarone and 8,9-dehydroxylarone also have cytotoxic activity. Several cytochalasins, compounds that bind to actin in muscle tissue, have been found in the fungus. *X. hypoxylon* also contains a carbohydrate-binding protein, a lectin, with unique sugar specificity and which has potent antitumor effects in various tumor cell lines (Fig. 4.2s) [557–559].

Streptomyces is the largest genus of Actinobacteria and the type genus of the family Streptomycetaceae. Found predominantly in soil and decaying vegetation, most streptomycetes produce spores and are noted for their distinct "earthy" odor that results from production of a volatile metabolite, geosmin. Streptomycetes are characterized by a complex secondary metabolism. They produce over two-thirds of the clinically useful antibiotics of natural origin (e.g., neomycin and chloramphenicol). *Streptomyces* is the largest antibiotic-producing genus, producing antibacterial, antifungal, and antiparasitic drugs and also a wide range of other bioactive compounds, such as immunosuppressants. Almost all of the bioactive compounds

Fig. 4.2s *Xylaria* sp.

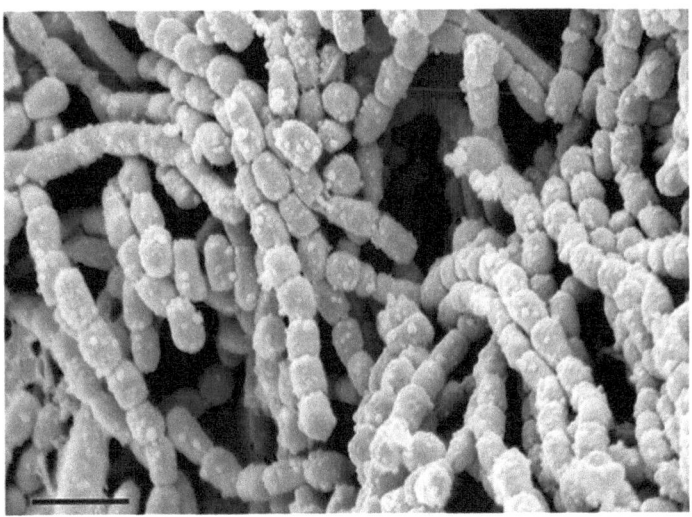

Fig. 4.2t *Streptomyces*

produced by *Streptomyces* are initiated during the time coinciding with the aerial hyphal formation from the substrate mycelium. The now uncommonly used streptomycin takes its name directly from *Streptomyces*. Streptomycetes are infrequent pathogens, though infections in humans, such as mycetoma can be caused by *S. somaliensis* and *S. sudanensis* and in plants can be caused by *S. caviscabies*, *S. acidiscabies*, *S. turgidiscabies* and *S. scabies* (Fig. 4.2t) [560, 561].

Zostera is a small genus of widely distributed sea grass, commonly called marine eelgrass or simply eelgrass. *Zostera japonica* is native to the seacoast of eastern Asia from Russia to Vietnam. *Z. japonica* inhabits previously unvegetated mid and high intertidal mudflats. *Z. japonica* exhibits both asexual (rhizomal growth) and sexual (seed production) reproduction. *Z. japonica* can also bind sediments creating a hummocked appearance compared to adjacent unvegetated mud and sand flats. *Zostera marina* is found on sandy substrates or in estuaries submerged or partially floating. Most *Zostera* are perennial. They have long, bright green, ribbon-like leaves, about 1 cm wide. Short stems grow up from extensive, white branching rhizomes. The flowers are enclosed in the sheaths of the leaf bases; the fruits are bladdery and can float. *Zostera* beds are important for sediment deposition, substrate stabilization, as substrate for epiphytic algae and micro-invertebrates, and as nursery grounds for many species of economically important fish and shellfish. It is an important food for brant geese and wigeons, and even caterpillars of the grass moth *Dolicharthria punctalis*. *Zostera* is able to maintain its turgor at a constant pressure in response to fluctuations in environmental osmolarity (Fig. 4.2u) [562, 563]

Tolypothrix is a genus of the tribe of cyanobacteria (blue-green algae). *Tolypothrix* is spurious branched, blue green to brown, and their cylindrical cells contain, like all bacterial cells, neither nuclei nor plastids. Each thread is in its own gelatinous sheath. An improper lateral branch is formed by fragmentation of a thread on a

Fig. 4.2u *Zostera* sp.

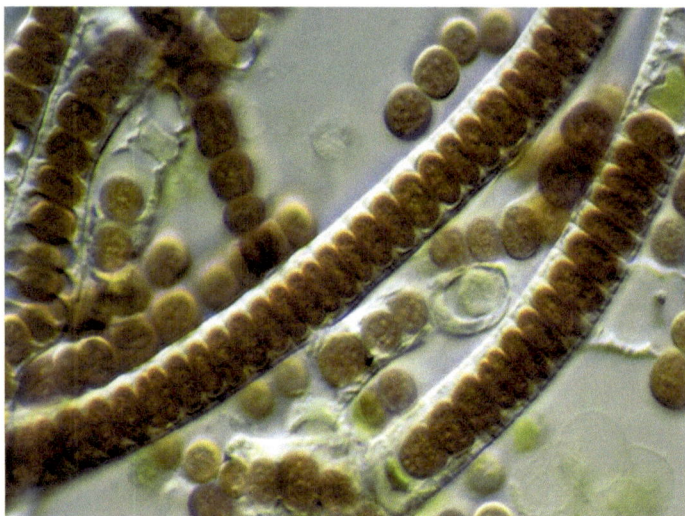

Fig. 4.2v *Tolypothrix nodosa*

preformed body, on which a single cell dies. In this way, two separate threads are formed which remains connected by gelatinous sheath. One of the threads breaks forth from the gelatinous sheath, the other thread with a varied range of the resulting branching point and one or more colorless, thick-walled cells called heterocysts. They are used for biological nitrogen fixation. Growth occurs by cell division within the thread (Fig. 4.2v).

Bursatella leachii, common name the ragged sea hare or shaggy sea hare, is a species of large sea slug or sea hare, a marine, gastropod mollusk in the family

Fig. 4.2w *Bursatella leachii*

Aplysiidae, the sea hares. *Bursatella leachii* is green to greenish brown. It has a broad and short head. Its mantle is covered with papillae (finger-like outgrowths), which give it a thorny aspect. The mantle has a network-like pattern with blue eye-spots (ocelli) in black spots and green areas. It moves slowly on a broad foot. There is short, sharp tail. The short parapodia (fleshy, winglike outgrowths) are fused on their rear end. The length is up to 15 cm but usually between 5 and 10 cm. The maximum recorded length is 120 mm (Fig. 4.2w) [564].

Nostoc commune is a species of cyanobacterium in the family Nostocaceae. Common names include star jelly, witch's butter, mare's eggs, and facai. It initially forms a small, hollow gelatinous globule which grows and becomes leathery, flattened, and convoluted, forming a gelatinous mass with other colonies growing nearby. Inside the thin sheath are numerous unbranched hair-like structures called trichomes formed of short cells in a string. Cells have no nucleus nor internal membrane system. Along the trichomes, larger specialist nitrogen-fixing cells called heterocysts occur between the ordinary cells. When wet, *Nostoc commune* is bluish green, olive green, or brown, but in dry conditions, it becomes an inconspicuous, crisp brownish mat. Nostoc commune does not have chloroplasts but contains photosynthetic pigments in the cytoplasm of the cells. It also contains pigments that absorb long and medium wavelength ultraviolet radiation, which enables it to survive in places with high levels of radiation. It has been found that extracellular polysaccharides are vital to its stress tolerance and ability to recover (Fig. 4.2x) [565].

Fig. 4.2x *Nostoc commune*

4.3   Animal Sources

It has been an oldest form of connection between mankind and nature. In this connection the use of animals in medical practices can be perceived as persistent relationship. The antiquity in the use of medicinal animals and its persistency through times are a testimony to the importance of those therapeutic resources to mankind (Table 4.3).

Python reticulatus is a species of python found in Southeast Asia. Adults can grow to 6.95 m in length but normally grow to an average of 3–6 m. They are the world's longest snakes and longest reptile, but are not the most heavily built. Like all pythons, they are nonvenomous constrictors and normally not considered dangerous to humans. Although large specimens are powerful enough to kill an adult human, attacks are only occasionally reported. An excellent swimmer, *Python reticulatus* has been reported far out at sea and has colonized many small islands within its range. The specific name, *reticulatus*, is Latin meaning net like, or reticulated, and is a reference to the complex color pattern. The color pattern is a complex geometric pattern that incorporates different colors. The back typically has a series of irregular diamond shapes flanked by smaller markings with light centers. In this species' wide geographic range, much variation of size, color, and markings commonly occurs. The smooth dorsal scales are arranged in 69–79 rows at midbody. There are deep pits on 4 anterior upper labials, on 2 or 3 anterior lower labials, and on 5 or 6 posterior lower labials (Fig. 4.3a) [572].

Table 4.3 Anti-inflammatory animal sources

Honey bees	Honey	Flavonoid	Antibacterial activity	[566]
Balaenoptera acutorostrata	Liver	n-3 fatty acids	Temperature modulation of oxygen transport	[567, 568]
Python reticulatus	Snake	Daboiatoxin	Antitoxic activity	[569]
Sparus aurata	Fish	Total fatty acids	Cytotoxic activity	[570, 571]

Fig. 4.3a *Python reticulates*

Sparus aurata is a fish of the bream family Sparidae found in the Mediterranean Sea and the eastern coastal regions of the North Atlantic Ocean. It commonly reaches about 35 cm in length, but may reach 70 cm and weight up to about 17 kg. The gilt-head bream is generally considered the best tasting of the breams. It is the single species of the genus *Sparus*. It mainly feeds on shellfish, but also some plant material (Fig. 4.3b).

Honey bees belong to the genus *Apis*, primarily distinguished by the production and storage of honey and the construction of perennial, colonial nests from wax. Honey bees represent only a small fraction of the roughly 20,000 known species of bees. All honey bees live in colonies where the workers sting intruders as a form of defense, and alarmed bees release a pheromone that stimulates the attack response in other bees. Bee pollen is rich in protein, particularly free amino acids, and also abounds with carbohydrate, lipid, vitamins, and minerals. In addition, bee pollen contains minor components, such as flavonoids and phenolic compounds. It is claimed to be effective for the treatment of asthma, bronchitis, cancers, peptic ulcers, colitis, various types of infections including hepatitis B, and rheumatism by the herb dealers in northeast Turkey (Fig. 4.3c) [573, 574].

Fig. 4.3b   *Sparus aurata*

Fig. 4.3c   *Apis mellifera*

Balaenoptera acutorostrata (minke whale) is a name given to marine mammal belonging to a clade within the suborder of baleen whales. *B. acutorostrata* are the second smallest whale measure an average of 6.9–7.4 m in length with 4–5 t of weight. Their life period is typically 30–50 years. It is black or purple in color. Minke whales have between 240 and 360 baleen plates on each side of their mouths. The tail extends into two long tips. The dorsal fin is high and curved back. Minke whales travel in small groups of 2–4 whales. They are thought to be curious, approaching ships and wharfs which is not typical of its family. Minkes are fast swimmers that are able to leap completely out of the water like a dolphin (Fig. 4.3d) [575].

Fig. 4.3d *Balaenoptera acutorostrata*

References

1. Rasmussen RS, Morrissey MT (2007) Marine biotechnology for production of food ingredients. Adv Food Nutr Res 52:237–292
2. Plaza M, Cifuentes A, Ibáñez E (2008) In the search of new functional food ingredients from algae. Trends Food Sci Technol 19:31–39
3. Aiello A, Borrelli F, Capasso R, Fattorusso E, Luciano P, Menna M (2003) Conicamin, a novel histamine antagonist from the mediterranean tunicate Aplidium conicum. Bioorg Med Chem Lett 13(24):4481–4483
4. Chao CH, Wen ZH, Wu YC, Yeh HC, Sheu JH (2008) Cytotoxic and anti-inflammatory cembranoids from the soft coral Lobophytum crassum. J Nat Prod 71(11):1819–1824. doi:10.1021/np8004584
5. Shen YC, Chen YH, Hwang TL, Guh JH, Khalil AT (2007) Four new briarane diterpenoids from the gorgonian coral Junceella fragilis. Helv Chim Acta 90(7):1391–1398
6. Kobayashi H, Kitamura K, Nagai K, Nakao Y, Fusetani N, van Soest RWM, Matsunaga S (2007) Carteramine A, an inhibitor of neutrophil chemotaxis, from the marine sponge Stylissa carteri. Tetrahedron Lett 48(12):2127–2129
7. Jayatilake GS, Freeberg DR, Liu Z, Richheimer SL, Blake ME, Bailey DT, Haridas V, Gutterman JU (2003) Isolation and structures of Avicins D and G: in vitro tumor-inhibitory Saponins derived from Acacia victoriae. J Nat Prod 66(6):779–783. doi:10.1021/np020400v
8. İşcan G, Kirimer N, Kürkçüoglu M, Arabaci T, Küpeli E, Başer KH (2006) Biological activity and composition of the essential oils of Achillea schischkinii Sosn. and Achillea aleppica DC. subsp. J Agric Food Chem 54(1):170–173. doi:10.1021/jf051644z
9. Küpeli E, Orhan İ, Küsmenoğlu Ş, Yeşilada E (2007) Evaluation of anti-inflammatory and antinociceptive activity of five Anatolian achillea species. Turk J Pharm Sci 4(2):89–99
10. De Caluwé E, Halamová K, Van Damme P (2009) Baobab (Adansonia digitata L.) a review of traditional uses, phytochemistry and pharmacology, African natural plant products. New discoveries and challenges in chemistry and quality. ACS Symp Ser 1021:51–84. doi:10.1021/bk-2009-1021.ch004
11. Yang X-W, Zhao J, Cui Y-X, Liu X-H, Ma C-M, Hattori M, Zhang L-H (1999) Anti-HIV-1 Protease Triterpenoid Saponins from the seeds of Aesculus chinensis. J Nat Prod 62(11):1510–1513. doi:10.1021/np990180u

12. Wei F, Ma SC, Ma LY, But PP, Lin RC, Khan IA (2004) Antiviral flavonoids from the seeds of Aesculus chinensis. J Nat Prod 67(4):650–653
13. Espín JC, Wichers HJ (1999) Slow-binding inhibition of mushroom (Agaricus bisporus) tyrosinase isoforms by tropolone. J Agric Food Chem 47(7):2638–2644. doi:10.1021/jf981055b
14. Chen S, Oh SR, Phung S, Hur G, Ye JJ, Kwok SL, Shrode GE, Belury M, Adams LS, Williams D (2006) Anti-aromatase activity of phytochemicals in white button mushrooms (Agaricus bisporus). Cancer Res 66(24):12026–12034
15. Wang Y-T, Huang Z-J, Chang H-M (2004) Proteomic analysis of human leukemic U937 cells incubated with conditioned medium of mononuclear cells stimulated by proteins from dietary mushroom of Agrocybe aegerita. J Proteome Res 3(4):890–896. doi:10.1021/pr049922h
16. Lo KM, Cheung PCK (2005) Antioxidant activity of extracts from the fruiting bodies of Agrocybe aegerita var. alba. Food Chem 89(4):533–539
17. Zhao C, Sun H, Tong X, Qi Y (2003) An antitumour lectin from the edible mushroom Agrocybe aegerita. Biochem J 374:321–327. doi:10.1042/BJ20030300
18. Takasaki M, Tokuda H, Nishino H, Konoshima T (1999) Cancer chemopreventive agents (Antitumor-promoters) from Ajuga decumbens. J Nat Prod 62(7):972–975. doi:10.1021/np990033w
19. Konoshima T, Takasaki M, Tokuda H, Nishino H (2000) Cancer chemopreventive activity of an iridoid glycoside, 8-acetylharpagide from Ajuga decumbens. Cancer Lett 157(1):87–92
20. Lawson LD (1998) Chapter 14: Garlic: a review of its medicinal effects and indicated active compounds. In: Phytomedicines of Europe, ACS symposium series 691. American Chemical Society, Washington, DC, pp 176–209. doi:10.1021/bk-1998-0691.ch014
21. Hutter JA, Salman M, Stavinoha WB, Satsangi N, Williams RF, Streeper RT, Weintraub ST (1996) Antiinflammatory C-Glucosyl Chromone from Aloe barbadensis. J Nat Prod 59(5):541–543. doi:10.1021/np9601519
22. Cock IE (2008) Antimicrobial activity of Aloe barbadensis miller leaf gel components. Int J Microbiol 4:2
23. Hu Y, Xu J, Hu Q (2003) Evaluation of antioxidant potential of aloe vera (Aloe barbadensis Miller) extracts. J Agric Food Chem 51(26):7788–7791. doi:10.1021/jf034255i
24. Ali MS, Tezuka Y, Awale S, Banskota AH, Kadota S (2001) Six sew Diarylheptanoids from the seeds of Alpinia blepharocaly. J Nat Prod 64(3):289–293. doi:10.1021/np000496y
25. Ali MS, Banskota AH, Tezuka Y, Saiki I, Kadota S (2001) Antiproliferative activity of diaryl-heptanoids from the seeds of Alpinia blepharocalyx. Biol Pharm Bull 24(5):525–528
26. Yang Y-L, Chang F-R, Wu C-C, Wang WY, Wu Y-C (2002) New ent-Kaurane diterpenoids with anti-platelet aggregation activity from Annona squamosa. J Nat Prod 65(10):1462–1467. doi:10.1021/np020191e
27. Kaleem M, Asif M, Ahmed QU, Bano B (2006) Antidiabetic and antioxidant activity of Annona squamosa extract in streptozotocin-induced diabetic rats. Singap Med J 47(8):670–675
28. Wu T-S, Chan Y-Y, Leu Y-L (2001) Constituents of the roots and stems of Aristolochia mollis-sima. J Nat Prod 64(1):71–74. doi:10.1021/np0002886
29. Yu JQ, Liao ZX, Cai XQ, Lei JC, Zou GL (2007) Composition, antimicrobial activity and cytotoxicity of essential oils from Aristolochia mollissima. Environ Toxicol Pharmacol 23(2):162–167. doi:10.1016/j.etap.2006.08.004
30. Shen C-C, Syu W-J, Li S-Y, Lin C-H, Lee G-H, Sun C-M (2002) Antimicrobial activities of naphthazarins from Arnebia euchroma. J Nat Prod 65(12):1857–1862. doi:10.1021/np010599w
31. Patil AD, Freyer AJ, Killmer L, Offen P, Taylor PB, Votta BJ, Johnson RK (2002) A new dimeric dihydrochalcone and a new prenylated flavone from the bud covers of artocarpus altilis: potent inhibitors of Cathepsin K. J Nat Prod 65(4):624–627. doi:10.1021/np0105634
32. Jain AP (2009) Evaluation of Anticonvulsant activity of methanolic extract of Artocarpus heterophyllus lam. (Moraceae) in mice. J Pharm Res 2(6):1004–1007
33. Khan MR, Omoloso AD, Kihara M (2003) Antibacterial activity of Artocarpus heterophyllus. Fitoterapia 74(5):501–505

34. Wei B-L, Weng J-R, Chiu P-H, Hung C-F, Wang J-P, Lin C-N (2005) Antiinflammatory flavonoids from Artocarpus heterophyllus and Artocarpus communis. J Agric Food Chem 53(10):3867–3871. doi:10.1021/jf047873n

35. Loizzo MR, Tundis R, Chandrika UG, Abeysekera AM, Menichini F, Frega NG (2010) Antioxidant and antibacterial activities on foodborne pathogens of Artocarpus heterophyllus Lam. (Moraceae) leaves extracts. J Food Sci 75(5):291–295

36. Lakheda S, Devalia R, Jain UK, Gupta N, Raghuwansi AS, Patidar N (2005) Antiinflammatory activity of Artocarpus heterophyllus bark, Pelagia Research Library. Der Pharmacia Sinica 2(2)

37. Jang DS, Cuendet M, Fong HHS, Pezzuto JM, Kinghorn AD (2004) Constituents of Asparagus officinalis evaluated for inhibitory activity against cyclooxygenase-2. J Agric Food Chem 52(8)):2218–2222. doi:10.1021/jf0305229

38. Huang XF, Lin YY, Kong LY (2008) Steroids from the roots of Asparagus officinalis and their cytotoxic activity. J Integr Plant Biol 50(6):717–722. doi:10.1111/j.1744-7909.2008.00651.x

39. Çalış İ, Zor M, Saracoğlu İ, Işımer A, Rüegger H (1996) Four novel cycloartane glycosides from Astragalus oleifolius. J Nat Prod 59(11):1019–1023. doi:10.1021/np9604184

40. Ozipek M, Dönmez AA, Caliş I, Brun R, Rüedi P, Tasdemir D (2005) Leishmanicidal cycloartane-type triterpene glycosides from Astragalus oleifolius. Phytochemistry 66(10):1168–1173

41. Resch M, Steigel A, Chen Z-l, Bauer R (1998) 5-lipoxygenase and cyclooxygenase-1 inhibitory active compounds from Atractylodes lancea. J Nat Prod 61(3):347–350. doi:10.1021/np970430b

42. Wang Y, Dai CC, Chen Y (2009) Antimicrobial activity of volatile oil from Atractylodes lancea against three species of endophytic fungi and seven species of exogenous fungi. J Appl Ecol 20(11):2778–2784

43. Lin Y, Jin T, Wu X, Huang Z, Fan J, Chan WL (1997) A novel bisesquiterpenoid, biatractylolide from the Chinese herbal plant Atractylodes macrocephala. J Nat Prod 60(1):27–28. doi:10.1021/np9603582

44. Jiang H, Shi J, Li Y (2011) Screening for compounds with aromatase inhibiting activities from Atractylodes macrocephala Koidz. Molecules 16(4):3146–3151

45. Bennett RN, Mellon FA, Rosa EAS, Perkins L, Kroon PA (2004) Profiling glucosinolates, flavonoids, alkaloids, and other secondary metabolites in tissues of Azima tetracantha L. (Salvadoraceae). J Agric Food Chem 52(19):5856–5862. doi:10.1021/jf040091+

46. Begum TN, Hussain M, Ilyas M, Anand AV (2011) Antipyretic activity of azima tetracantha in experimental animals. Int J Curr Biomed Pharm Res 1(2):41–44

47. Nakasugi T, Komai K (1998) Antimutagens in the Brazilian folk medicinal plant Carqueja (Baccharis trimera Less.). J Agric Food Chem 46(7):2560–2564. doi:10.1021/jf9711045

48. Soicke H, Leng Peschlow E (1987) Characterization of flavonoids from baccharis trimera and their antihepatotoxic properties. Planta Med 53(1):37–39. doi:10.1055/s-2006-962613

49. Zhou Y, Shen Y-H, Zhang C, Su J, Liu R-H, Zhang W-D (2007) Triterpene saponins from bacopa monnieri and their antidepressant effects in two mice models. J Nat Prod 70(4):652–655. doi:10.1021/np060470s

50. Ghosh T, Maity TK, Singh J (2011) Antihyperglycemic activity of bacosine, a triterpene from Bacopa monnieri, in alloxan-induced diabetic rats. Planta Med 77(8):804–808

51. Braca A, De Tommasi N, Di Bari L, Pizza C, Politi M, Morelli I (2001) Antioxidant principles from Bauhinia tarapotensis. J Nat Prod 64(7):892–895. doi:10.1021/np0100845

52. Ju J-H, Liu D, Lin G, Xu D, Han B, Yang J-s, Tu G-z, Ma L-b (2002) Beesiosides a–F, six new cycloartane triterpene glycosides from Beesia calthaefolia. J Nat Prod 65(1):42–47. doi:10.1021/np010293p

53. Mazumder S, Morvan C, Thakur S, Ray B (2004) Cell wall polysaccharides from chalkumra (Benincasa hispida) fruit. Part I. Isolation and characterization of pectins. J Agric Food Chem 52(11):3556–3562. doi:10.1021/jf0343130

54. Samad NB, Debnath T, Jin HL, Lee BR, Park PJ, Lee SY, Lim BO (June) Antioxidant Activity of Benincasa Hispida Seeds. J Food Biochem, doi:10.1111/J.1745-4514.2011.00643.X

55. De Tommasi N, Aquino R, Cumandà J, Mahmood N (1997) Flavonol and chalcone ester glycosides from Bidens leucantha. J Nat Prod 60(3):270–273. doi:10.1021/np960572q
56. Inngjerdingen KT, Coulibaly A, Diallo D, Michaelsen TE, Paulsen BS (2006) A complement fixing polysaccharide from biophytum petersianum Klotzsch, a medicinal plant from Mali, West Africa. Biomacromolecules 7(1):48–53. doi:10.1021/bm050330h
57. Inngjerdingen M, Inngjerdingen KT, Patel TR, Allen S, Chen X, Rolstad B, Morris GA, Harding SE, Michaelsen TE, Diallo D, Paulsen BS (2008) Pectic polysaccharides from Biophytum petersianum Klotzsch, and their activation of macrophages and dendritic cells. Glycobiology 18(12):1074–1084. doi:10.1093/glycob/cwn090
58. Takashima J, Ohsaki A (2001) Acutifolins A-F, a New flavan-derived constituent and five new flavans from Brosimum acutifolium. J Nat Prod 64(12):1493–1496. doi:10.1021/np010389j
59. Takashima J, Komiyama K, Ishiyama H, Kobayashi J, Ohsaki A (2005) Brosimacutins J-M, four new flavonoids from Brosimum acutifolium and their cytotoxic activity. Planta Med 71(7):654–658
60. Cuendet M, Pezzuto JM (2004) Antitumor activity of bruceantin: an old drug with new promise. J Nat Prod 67(2):269–272. doi:10.1021/np030304+
61. Gillin FD, Reiner DS, Suffness M (1982) Bruceantin, a potent amoebicide from a plant, Brucea antidysenterica. Antimicrob Agents Chemother 22(2):342–345
62. Liao Y-H, Houghton PJ, Hoult JRS (1999) Novel and known constituents from Buddleja species and their activity against leukocyte eicosanoid generation. J Nat Prod 62(9):1241–1245. doi:10.1021/np990092+
63. Houghton PJ, Hikino H (1989) Anti-hepatotoxic activity of extracts and constituents of Buddleja species. Planta Med 55(2):123–126
64. Hernández-Hernández JD, Román-Marín LU, Cerda-García-Rojas CM, Joseph-Nathan P (2005) Verticillane derivatives from Bursera suntui and Bursera kerberi. J Nat Prod 68(11):1598–1602. doi:10.1021/np050323e
65. García-Gutiérrez HA, Cerda-García-Rojas CM, Hernández-Hernández JD, Román-Marín LU, Joseph-Nathana P (2008) Oxygenated verticillene derivatives from Bursera suntui. Phytochemistry 69:2844–2848
66. Kalauni SK, Awale S, Tezuka Y, Banskota AH, Linn TZ, Kadota S (2004) Cassane- and nor-cassane-type diterpenes of Caesalpinia crista from Myanmar. J Nat Prod 67(11):1859–1863. doi:10.1021/np049742m
67. Kalauni SK, Awale S, Tezuka Y, Banskota AH, Linn TZ, Asih PB, Syafruddin D, Kadota S (2006) Antimalarial activity of cassane- and norcassane-type diterpenes from Caesalpinia crista and their structure-activity relationship. Biol Pharm Bull 29(5):1050–1052
68. Lee C-P, Yen G-C (2006) Antioxidant activity and bioactive compounds of tea seed (Camellia oleifera Abel.) oil. J Agric Food Chem 54(3):779–784. doi:10.1021/jf052325a
69. Ramji D, Sang S, Liu Y, Rosen RT, Ghai G, Ho C-T, Yang CS, Huang M-T (2005) Chapter 20: Effect of black tea theaflavins and related benzotropolone derivatives on 12-O-tetradecanoylphorbol-13-acetate-induced mouse ear inflammation and inflammatory mediators. In: Fereidoon S, Chi-Tang H (eds) Phenolic compounds in foods and natural health products, ACS symposium series, 909. American Chemical Society, Washington, DC, pp 242–253. doi:10.1021/bk-2005-0909.ch020
70. Henning SM, Niu Y, Lee NH, Thames GD, Minutti RR, Wang H, Go VL, Heber D (2004) Bioavailability and antioxidant activity of tea flavanols after consumption of green tea, black tea, or a green tea extract supplement. Am J Clin Nutr 80(6):1558–1564
71. Jhoo J-W (2007) Antioxidant and anti-cancer activities of green and black tea polyphenols, antioxidant measurement and applications. ACS Symp Ser 956:215–225. doi:10.1021/bk-2007-0956.ch015
72. Viegas C Jr, Bolzani V d S, Furlan M, Furlan M, Barreiro EJ, Young MCM, Tomazela D, Eberlin MN (2004) Further bioactive piperidine alkaloids from the flowers and green fruits of Cassia spectabilis. J Nat Prod 67(5):908–910. doi:10.1021/np0303963

73. Torey A, Sasidharan S (2011) Anti-Candida albicans biofilm activity by Cassia spectabilis standardized methanol extract: an ultrastructural study. Eur Rev Med Pharmacol Sci 15(8):875–882

74. Yen G-C, Chuang D-Y (2000) Antioxidant properties of water extracts from Cassia tora L. In relation to the degree of roasting. J Agric Food Chem 48(7):2760–2765. doi:10.1021/jf991010q

75. Yang C, Yuan C, Jia Z (2003) Xanthanolides, germacranolides, and other constituents from Carpesium longifolium. J Nat Prod 66(12):1554–1557. doi:10.1021/np030278f

76. Yang C, Yuan C, Jia Z (2003) Xanthanolides, germacranolides, and other constituents from Carpesium longifolium. J Nat Prod 66(12):1554–1557

77. Matsuo Y, Watanabe K, Mimaki Y (2009) Triterpene glycosides from the underground parts of Caulophyllum thalictroides. J Nat Prod 72(6):1155–1160

78. Koorbanally NA, Randrianarivelojosia M, Mulholland DA, Quarles van Ufford L, van den Berg AJ (2002) Bioactive constituents of Cedrelopsis microfoliata. J Nat Prod 65(9):1349–1352

79. Hwang BY, Kim HS, Lee JH, Hong YS, Ro JS, Lee KS, Lee JJ (2001) Antioxidant benzoylated flavan-3-ol glycoside from Celastrus orbiculatus. J Nat Prod 64(1):82–84. doi:10.1021/np0002511

80. Park HJ, Cha DS, Jeon H (2011) Antinociceptive and hypnotic properties of Celastrus orbiculatus. J Ethnopharmacol 137(3):1240–1244

81. Dini I, Tenore GC, Dini A (2002) Oleanane saponins in kancolla, a sweet variety of Chenopodium quinoa. J Nat Prod 65(7):1023–1026. doi:10.1021/np010625q

82. Chao LK, Hua K-F, Hsu H-Y, Cheng S-S, Liu J-Y, Chang S-T (2005) Study on the antiinflammatory activity of essential oil from leaves of Cinnamomum osmophloeum. J Agric Food Chem 53(18):7274–7278. doi:10.1021/jf051151u

83. Yoo KM, Lee KW, Park JB, Lee HJ, Hwang IK (2004) Variation in major antioxidants and total antioxidant activity of Yuzu (Citrus junos Sieb ex Tanaka) during maturation and between cultivars. J Agric Food Chem 52(19):5907–5913. doi:10.1021/jf0498158

84. Ho S-C, Lin C-C (2008) Investigation of heat treating conditions for enhancing the anti-inflammatory activity of citrus fruit (Citrus reticulata) peels. J Agric Food Chem 56(17):7976–7982. doi:10.1021/jf801434c

85. Jayaprakasha GK, Negi PS, Sikder S, Rao LJ, Sakariah KK (2000) Antibacterial activity of Citrus reticulata peel extracts. Z Naturforsch C 55(11–12):1030–1034

86. Li S, Lo C-Y, Ho C-T (2006) Hydroxylated polymethoxyflavones and methylated flavonoids in sweet orange (Citrus sinensis) peel. J Agric Food Chem 54(12):4176–4185. doi:10.1021/jf060234n

87. Li S, Lo C-Y, Dushenkov S, Ho C-T (2008) Polymethoxyflavones: chemistry, biological activity and occurrence in orange peel. In: Dietary supplements, ACS symposium series 987. American Chemical Society, Washington, DC, pp 191–210. doi:10.1021/bk-2008-0987.ch013

88. Anagnostopoulou MA, Kefalas P, Papageorgiou VP, Assimopoulou AN, Boskou D (2006) Radical scavenging activity of various extracts and fractions of sweet orange peel (Citrussinensis). Food Chem 94(1):19–25

89. Kuo C-C, Chiang W, Liu G-P, Chien Y-L, Chang J-Y, Lee C-K, Lo J-M, Huang S-L, Shih M-C, Kuo Y-H (2002) 2,2'-Diphenyl −1-picrylhydrazyl radical-scavenging active components from adlay (Coix lachryma-jobi L. Var. ma-yuen Stapf) hulls. J Agric Food Chem 50(21):5850–5855. doi:10.1021/jf020391w

90. Su Y, Guo D, Guo H, Liu J, Zheng J, Koike K, Nikaido T (2001) Four new triterpenoid saponins from Conyza blinii. J Nat Prod 64(1):32–36. doi:10.1021/np000310v

91. Lee H-S (2002) Rat lens aldose reductase inhibitory activities of coptis japonica root-derived isoquinoline alkaloids. J Agric Food Chem 50(24):7013–7016. doi:10.1021/jf020674o

92. Kim JP, Jung MY, Kim JP, Kim SY (2000) Antiphotooxidative activity of protoberberines derived from Coptis japonica makino in the chlorophyll-sensitized photooxidation of oil. J Agric Food Chem 48(4):1058–1063

93. Cho JY, Baik KU, Yoo ES, Yoshikawa K, Park MH (2006) In vitro antiinflammatory effects of neolignan woorenosides from the rhizomes of Coptis japonica. J Nat Prod 63(9):1205–1209. doi:10.1021/np9902791
94. Abdel-Halim OB, Morikawa T, Ando S, Matsuda H, Yoshikawa M (2004) New crinine-type alkaloids with inhibitory effect on induction of inducible nitric oxide synthase from Crinum yemense. J Nat Prod 67(7):1119–1124. doi:10.1021/np030529k
95. Abdel-Halim OB, Marzouk AM, Mothana R, Awadh N (2008) A new tyrosinase inhibitor from Crinum yemense as potential treatment for hyperpigmentation. Pharmazie 63(5):405–407
96. Fattorusso E, Taglialatela-Scafati O, Campagnuolo C, Santelia FU, Appendino G, Spagliardi P (2002) Diterpenoids from Cascarilla (Croton eluteria Bennet). J Agric Food Chem 50(18):5131–5138. doi:10.1021/jf0203693
97. Nath R, Roy S, De B, Dutta CM (2013) Anticancer and antioxidant activity of croton: a review. Int J Pharmacy Pharm Sci 5(2):63–70
98. Kuo P-C, Shen Y-C, Yang M-L, Wang S-H, Dinh TT, Dung NX, Chiang P-C, Lee K-H, Lee E-J, Wu T-S (2007) Crotonkinins A and B and related diterpenoids from Croton tonkinensis as anti-inflammatory and antitumor agents. J Nat Prod 70(12):1906–1909. doi:10.1021/np070383f
99. Giang PM, Son PT, Matsunami K, Otsuka H (2006) Anti-staphylococcal activity of ent-kaurane-type diterpenoids from Croton tonkinensis. J Nat Med 60:93–95
100. Pande M, Dubey VK, Yadav SC, Jagannadham MV (2006) A novel serine protease cryptolepain from Cryptolepis buchanani: purification and biochemical characterization. J Agric Food Chem 54(26):10141–10150. doi:10.1021/jf062206a
101. Vinayaka KS, Prashith KTR, Mallikarjun N, Sateesh VN (2010) Anti-dermatophyte activity of Cryptolepis buchanani Roem. & Schult. Pharmacognosy J 2(7):170–172
102. Bierer DE, Fort DM, Mendez CD et al (1998) Ethnobotanical-directed discovery of the anti-hyperglycemic properties of cryptolepine: its isolation from Cryptolepis sanguinolenta, synthesis, and in vitro and in vivo activities. J Med Chem 41(6):894–901. doi:10.1021/jm9704816
103. Wichtl M (1998) Curcuma (tumeric): biological activity and active compounds. In: Phytomedicines of Europe chemistry and biological activity, ACS symposium series. American Chemical Society, Washington, DC, pp 133–139. doi:10.1021/bk-1998-0691. ch011
104. Gupta SK, Agarwal R, Srivastava S, Agarwal P, Agrawal SS, Saxena R, Galpalli N (2008) The anti-inflammatory effects of Curcuma longa and berberis aristata in endotoxin-induced uveitis in rabbits. Invest Ophthalmol Vis Sci 49(9):4036–4040. doi:10.1167/iovs.07-1186
105. Nishiyama T, Mae T, Kishida H, Tsukagawa M, Mimaki Y, Kuroda M, Sashida Y, Takahashi K, Kawada T, Nakagawa K, Kitahara M (2005) Curcuminoids and sesquiterpenoids in turmeric (Curcuma longa L.) suppress an increase in blood glucose level in type 2 Diabetic KK-A^{y}Mice. J Agric Food Chem 53(4):959–963. doi:10.1021/jf0483873
106. Liju VB, Jeena K, Kuttan R (2011) An evaluation of antioxidant, anti-inflammatory, and antinociceptive activities of essential oil from Curcuma longa L. Indian J Pharm 43(5):526–531. doi:10.4103/0253-7613.84961
107. De Tommasi N, De Simone F, Speranza G, Pizza C (1996) Studies on the constituents of Cyclanthera pedata (Caigua) seeds: isolation and characterization of six new cucurbitacin glycosides. J Agric Food Chem 44(8):2020–2025. doi:10.1021/jf950532c
108. Cheel J, Theoduloz C, Rodríguez J, Schmeda-Hirschmann G (2005) Free radical scavengers and antioxidants from lemongrass (Cymbopogon citratus (DC.) Stapf.). Food Chem 53(7):2511–2517. doi:10.1021/jf0479766
109. El Bitar H, Van Nguyen H, Gramain A, Sévenet T, Bodo B (2004) New Alkaloids from Daphniphyllum calycinum. J Nat Prod 67(7):1094–1099. doi:10.1021/np040038f
110. Gamez EJ, Luyengi L, Lee SK, Zhu LF, Zhou BN, Fong HH, Pezzuto JM, Kinghorn AD (1998) Antioxidant flavonoid glycosides from Daphniphyllum calycinum. J Nat Prod 61(5):706–708
111. Lin T-H, Chang S-J, Chen C-C, Wang J-P, Tsao L-T (2001) Two phenanthraquinones from Dendrobium moniliforme. J Nat Prod 64(8):1084–1086. doi:10.1021/np010016i

112. Shu W, Fengjuan W, Yongping C (2009) Anti-oxidation activity in vitro of polysaccharides of Dendrobium Huoshanense and Dendrobium moniliforme, Institute of Agricultural Information, Chinese Academy of Agricultural Sciences; http://agris.fao.org/aos/records/CN2010001368

113. Lin YM, Anderson H, Flavin MT, Pai YH, Mata-Greenwood E, Pengsuparp T, Pezzuto JM, Schinazi RF, Hughes SH, Chen FC (1997) In vitro anti-HIV activity of biflavonoids isolated from Rhus succedanea and Garcinia multiflora. J Nat Prod 60(9):884–888

114. Su BN, Park EJ, Nikolic D, Schunke Vigo J, Graham JG, Cabieses F, van Breemen RB, Fong HH, Farnsworth NR, Pezzuto JM, Kinghorn AD (2003) Activity-guided isolation of novel norwithanolides from Deprea subtriflora with potential cancer chemopreventive activity. J Org Chem 68(6):2350–2361. doi:10.1021/jo020542u

115. Johansson S, Göransson U, Luijendijk T, Backlund A, Claeson P, Bohlin L (2002) A neutrophil multitarget functional bioassay to detect anti-inflammatory natural products. J Nat Prod 65(1):32–41. doi:10.1021/np010323o

116. Lin AM-Y, Wu L-Y, Hung K-C, Huang H-J, Lei YP, Lu W-C, Hwang LS (2012) Neuroprotective effects of longan (Dimocarpus longan lour.) flower water extract on MPP+–induced neurotoxicity in rat brain. J Agric Food Chem 60(36):9188–9194. doi:10.1021/jf302792

117. Zheng G-m, Xu L-x, Xie H-h, WU P, Wei X-y (2010) Chemical constituents from the pulps of Dimocarpus longan. J Trop Subtrop Bot 1:82–86

118. De Souza NJ (1993) Rohitukine and forskolin second-generation immunomodulatory, intraocular-pressure-lowering, and cardiotonic analogues. Hum Med Agents Plant 534:331–340. doi:10.1021/bk-1993-0534.ch022

119. Lawson LD, Bauer R (1998) Echinacea: biological effects and active principles. In: Phytomedicines of Europe, ACS symposium series, 691. American Chemical Society, Washington, DC, pp 140–157. doi:10.1021/bk-1998-0691.ch012

120. Watanabe M (1999) Antioxidative phenolic compounds from Japanese Barnyard Millet (Echinochloa utilis) grains. J Agric Food Chem 47(11):4500–4505. doi:10.1021/jf990498s

121. Yokozawa T, Kim HY, Kim HJ et al (2007) Amla (Emblica officinalis Gaertn.) attenuates age-related renal dysfunction by oxidative stress. J Agric Food Chem 55(19):7744–7752. doi:10.1021/jf072105s

122. Jeena JK, Kuttan R (2000) Hepatoprotective activity of Emblica officinalis and Chyavanaprash. J Ethnopharmacol 72(2):135–140

123. Asmawi MZ, Kankaanranta H, Moilanen E, Vapaatalo H (1993) Anti-inflammatory activities of Emblica officinalis Gaertn leaf extracts. J Pharm Pharmacol 45(6):581–584

124. Patel SS, Goyal RK (2012) Emblica officinalis Geart: a comprehensive review on phytochemistry, pharmacology and ethnomedicinal uses. Res J Med Plant 6:6–16. doi:10.3923/rjmp.2012.6.16

125. Shikov AN, Poltanov EA, Damien Dorman HJ, Makarov VG, Tikhono VP, Hiltunen R (2006) Chemical composition and in vitro antioxidant evaluation of commercial water-soluble Willow Herb (Epilobium angustifolium L.) extracts. J Agric Food Chem 54(10):3617–3624. doi:10.1021/jf052606i

126. Le Claire E, Schwaiger S, Banaigs B, Stuppner H, Gafner F (2005) Distribution of a new rosmarinic acid derivative in Eryngium alpinum L. and other Apiaceae. J Agric Food Chem 53(11):4367–4372. doi:10.1021/jf050024v

127. Miyazawa M, Hisama M (2003) Antimutagenic activity of phenylpropanoids from clove (Syzygium aromaticum). J Agric Food Chem 51(22):6413–6422. doi:10.1021/jf030247q

128. Patel BK, Jagannadham MV (2003) A high cysteine containing thiol proteinase from the latex of Ervatamia heyneana: purification and comparison with Ervatamin B and C from Ervatamia coronaria. J Agric Food Chem 51(21):6326–6334. doi:10.1021/jf026184d

129. Silva GL, Cui B, Chávez D et al (2001) Modulation of the multidrug-resistance phenotype by new tropane alkaloid aromatic esters from Erythroxylum pervillei. J Nat Prod 64(12):1514–1520. doi:10.1021/np010295+

130. Agarwal N, Chandra A, Tyagi LK (2011) Herbal medicine: alternative treatment for cancer therapy. Int J Pharma Sci Res 2(9):2249–2258

131. Hegde VR, Dai P, Patel MG, Puar MS, Das P, Pai J, Bryant R, Cox PA (1997) Phospholipase A$_2$ inhibitors from an erythrina species from Samoa. J Nat Prod 60(6):537–539. doi:10.1021/np960533e

132. Tanaka H, Sato M, Fujiwara S, Hirata M, Etoh H, Takeuchi H (2002) Antibacterial activity of isoflavonoids isolated from Erythrina variegata against methicillin-resistant Staphylococcus aureus. Lett Appl Microbiol 35(6):494–498

133. Damu AG, Kuo P-C, Shi L-S, Li C-Y, Kuoh C-S, Wu P-L, Wu T-S (2005) Phenanthroindolizidine alkaloids from the stems of Ficus septic. J Nat Prod 68(7):1071–1075. doi:10.1021/np050095o

134. Nugroho AE, Hermawan A, Nastiti K, Suven, Elisa P, Hadibarata T, Meiyanto E (2012) Immunomodulatory effects of hexane insoluble fraction of Ficus septica Burm. F. in doxorubicin-treated rats. Asian Pac J Cancer Prev 13(11):5785–5790

135. Yoshikawa K, Inoue M, Matsumoto Y, Sakakibara C, Miyataka H, Matsumoto H, Arihara S (2005) Lanostane triterpenoids and triterpene glycosides from the fruit body of Fomitopsis pinicola and their inhibitory activity against COX-1 and COX-2. J Nat Prod 68(1):69–73. doi:10.1021/np040130b

136. Choi DB, Park S-S, Ding J-L, Cha W-S (2007) Effects of Fomitopsis pinicola extracts on antioxidant and antitumor activities. Biotechnol Bioprocess Eng 12(5):516–524

137. Wang SY, Bunce JA, Maas JL (2003) Elevated carbon dioxide increases contents of antioxidant compounds in field-grown strawberries. J Agric Food Chem 51(15):4315–4432. doi:10.1021/jf021172d

138. Kleinwächter P, Anh N, Kiet TT, Schlegel B, Dahse H-M, Härtl A, Gräfe U (2001) Colossolactones, new triterpenoid metabolites from a Vietnamese mushroom Ganoderma colossum. J Nat Prod 64(2):236–239. doi:10.1021/np000437k

139. Ofodile LN, Uma N, Grayer RJ, Ogundipe OT, Simmonds MS (2012) Antibacterial compounds from the mushroom Ganoderma colossum from Nigeria. Phytother Res 26(5):748–751. doi:10.1002/ptr.3598

140. Huang M-T, Liu Y, Badmaev V, Ho C-T (2008) Dietary supplements. In: Antiinflammatory and anticancer activities of garcinol, ACS symposium series 987, pp 293–303. doi:10.1021/bk-2008-0987.ch020

141. Yamaguchi F, Ariga T, Yoshimura Y, Nakazawa H (2000) Antioxidative and anti-glycation activity of garcinol from Garcinia indica fruit rind. J Agric Food Chem 48(2):180–185. doi:10.1021/jf990845y

142. Khatib NA, Pawase K, Patil PA (2010) Evaluation of antiinflammation activity of Garcinia Indica fruit rind extracts in Wistar rats. Int J Res Ayur Pharm 1(2):449–454

143. Huang Y-L, Chen C-C, Chen Y-J, Huang R-L, Shieh B-J (2001) Three Xanthones and a Benzophenone from Garcinia mangostana. J Nat Prod 64(7):903–906. doi:10.1021/np000583q

144. Suksamrarn S, Suwannapoch N, Phakhodee W, Thanuhiranlert J, Ratananukul P, Chimnoi N, Suksamrarn A (2003) Antimycobacterial activity of prenylated xanthones from the fruits of Garcinia mangostana. Chem Pharm Bull 51(7):857–859

145. Chen L-G, Yang L-L, Wang C-C (2008) Anti-inflammatory activity of mangostins from Garcinia mangostana. Food Chem Toxicol 46:688–693

146. Chiang Y-M, Kuo Y-H, Oota S, Fukuyama Y (2003) Xanthones and benzophenones from the Stems of Garcinia multiflora. J Nat Prod 66(8):1070–1073. doi:10.1021/np030065q

147. Lee JH, Lee DU, Jeong CS (2009) Gardenia jasminoides Ellis ethanol extract and its constituents reduce the risks of gastritis and reverse gastric lesions in rats. Food Chem Toxicol 47(6):1127–11231

148. Lin K-W, Huang A-M, Tu H-Y, Lee L-Y, Wu C-C, Hour T-C, Yang S-C, Pu Y-S, Lin C-N (2011) Xanthine oxidase inhibitory triterpenoid and phloroglucinol from guttiferaceous plants inhibit growth and induced apoptosis in human NTUB1 cells through a ROS-dependent mechanism. J Agric Food Chem 59(1):407–414. doi:10.1021/jf1041382

149. Minami H, Takahashi E, Fukuyama Y, Kodama M, Yoshizawa T, Nakagawa K (1995) Novel xanthones with superoxide scavenging activity from Garcinia subelliptica. Chem Pharm Bull 43(2):347–349

150. Maldonado PD, Rivero-Cruz I, Mata R, Pedraza-Chaverr J (2005) Antioxidant activity of A-Type Proanthocyanidins from Geranium niveum (Geraniaceae). J Agric Food Chem 53(6):1996–2001. doi:10.1021/jf0483725
151. Şöhretoğlu D, Ekizoğlu M, Özalp M (2008) Free radical scavenging and antimicrobial activities of some geranium species. J Fac Pharm 28(2):115–124
152. Biondi DM, Rocco C, Ruberto G (2003) New dihydrostilbene derivatives from the leaves of Glycyrrhiza glabra and evaluation of their antioxidant activity. J Nat Prod 66(4):477–480. doi:10.1021/np020365s
153. Ambawade SD, Kasture VS, Kasture SB (2002) Anticonvulsant activity of roots and rhizomes of Glycyrrhiza glabra. Indian J Pharmacol Short Commun 34(4):251–255
154. Tanaka A, Shibamoto T (2008) Antioxidant and antiinflammatory activities of Licorice Root (Glycyrrhiza uralensis), aroma extract. In: Functional food and health, ACS symposium series 993, pp 229–237. doi:10.1021/bk-2008-0993.ch020
155. Zhang J, Gao WY, Yan S, Zhao Y (2012) Effects of space flight on the chemical constituents and anti-inflammatory activity of licorice (Glycyrrhiza uralensis Fisch). Int J Prod Res 11(2):601–609
156. Yao CS, Lin M, Wang L (2006) Isolation and biomimetic synthesis of anti-inflammatory stilbenolignans from Gnetum cleistostachyum. Chem Pharm Bull 54(7):1053–1057
157. Cheng K-W, Wang M, Chen F, Ho C-T (2008) Dietary supplements, oligostilbenes from gnetum species and anticarcinogenic and antiinflammatory activities of oligostilbenes, ACS symposium series 987: 36–58, doi:10.1021/bk-2008-0987.ch004
158. Beg S, Swain S, Hasan H et al (2011) Systematic review of herbals as potential anti-inflammatory agents: recent advances, current clinical status and future perspectives. Pharmacogn Rev 5(10):120–137. doi.10.4103/0973-7847.91102
159. Seger C, Godejohann M, Tseng L-H et al (2005) LC-DAD-MS/SPE-NMR hyphenation. A tool for the analysis of pharmaceutically used plant extracts: identification of Isobaric Irid Glycoside Regioisomers from Harpagophytum procumbens. Anal Chem 77(3):878–885. doi:10.1021/ac048772r
160. Mahomed IM, Ojewole JA (2006) Anticonvulsant activity of Harpagophytum procumbens DC [Pedaliaceae] secondary root aqueous extract in mice. Brain Res Bull 69(1):57–62
161. Kim Y, Park EJ, Kim J et al (2001) Neuroprotective constituents from Hedyotis diffusa. J Nat Prod 64(1):75–78. doi:10.1021/np000327d
162. Lin J, Chen Y, Wei L, Chen X, Xu W, Hong Z, Sferra TJ, Peng J (2010) Hedyotis Diffusa Willd extract induces apoptosis via activation of the mitochondrion-dependent pathway in human colon carcinoma cells. Int J Oncol 37(5):1331–1338
163. Kraus CM, Neszmélyi A, Holly S et al (1998) New acetylenes isolated from the bark of Heisteria acuminate. J Nat Prod 61(4):422–427. doi:10.1021/np970357p
164. Ukiya M, Akihisa T, Tokuda H (2003) Isolation, structural elucidation, and inhibitory effects of terpenoid and lipid constituents from sunflower pollen on Epstein – Barr virus early antigen induced by tumor promoter, TPA. J Agric Food Chem 51(10):2949–2957. doi:10.1021/jf0211231
165. Sala A, Recio M d C, Giner RM et al (2001) New acetophenone glucosides isolated from extracts of Helichrysum italicum with antiinflammatory activity. J Nat Prod 64(10):1360–1362. doi:10.1021/np010125x
166. Sala A, Recio M, Giner RM, Máñez S, Tournier H, Schinella G, Ríos JL (2002) Anti-inflammatory and antioxidant properties of Helichrysum italicum. J Pharm Pharmacol 54(3):365–371
167. Delgados G, del Socorro Olivares M, Chávez MI et al (2001) Antiinflammatory constituents from Heterotheca inuloides. J Nat Prod 64(7):861–864. doi:10.1021/np0005107
168. Coballase-Urrutia E, Pedraza-Chaverri J, Camacho-Carranza R (2010) Antioxidant activity of Heterotheca inuloides extracts and of some of its metabolites. Toxicology 276(1):41–48
169. Monteiro R, Becker H, Azevedo I et al (2006) Effect of Hop (Humulus lupulus L.) flavonoids on aromatase (Estrogen Synthase) activity. J Agric Food Chem 54(8):2938–2943. doi:10.1021/jf053162t

170. Zanoli P, Rivasi M, Zavatti M, Brusiani F, Baraldi M (2005) New insight in the neuropharma-
cological activity of Humulus lupulus L. J Ethnopharmacol 102(1):102–106
171. Pettit GR, Meng Y, Stevenson CA et al (2003) Isolation and structure of palstatin from the
Amazon tree Hymeneae palustris. J Nat Prod 66(2):259–262. doi:10.1021/np020231e
172. Kim SH, Sung SH, Choi SY et al (2005) Idesolide: a new spiro compound from Idesia polycarpa. Org Lett 7(15):3275–3277. doi:10.1021/ol051105f
173. Hwang JH, Moon SA, Lee CH et al (2012) Idesolide inhibits the adipogenic differentiation
of mesenchymal cells through the suppression of nitric oxide production. Eur J Pharmacol
685(1–3):218–223
174. Chou C-J, Lin L-C, Tsai W-J et al (1997) Phenyl β-D-glucopyranoside derivatives from the
fruits of Idesia polycarpa. J Nat Prod 60(4):375–377. doi:10.1021/np960335n
175. Esmelindro ÂA, Girardi JDS, Mossi A et al (2004) Influence of agronomic variables on the
composition of mate tea leaves (ilex paraguariensis) extracts obtained from CO_2 extraction at
30 °C and 175 bar. J Agric Food Chem 52(7):1990–1995. doi:10.1021/jf035143u
176. Burris KP, Davidson PM, Stewart CN et al (2011) Antimicrobial activity of Yerba Mate (Ilex
paraguariensis) aqueous extracts against Escherichia coli O157:H7 and Staphylococcus
aureus. J Food Sci 76(6):456–462. doi:10.1111/j.1750-3841.2011.02255.x
177. Abu Zarga MH, Hamed EM, Sabri SS et al (1998) New sesquiterpenoids from the Jordanian
medicinal plant Inula viscose. J Nat Prod 61(6):798–800. doi:10.1021/np9701992
178. Cafarchia C, De Laurentis N, Milillo MA et al (2002) Antifungal activity of essential oils
from leaves and flowers of Inula viscosa (Asteraceae) by Apulian region. Parassitologia
44(3–4):153–156
179. Hernández V, Del Recio M, Carmen MS et al (2001) A mechanistic approach to the in vivo
anti-inflammatory activity of sesquiterpenoid compounds isolated from Inula viscosa. Planta
Med 67(8):726–731
180. Máñez S, Recio M d C, Gil I et al (1999) A glycosyl analogue of diacylglycerol and other
antiinflammatory constituents from Inula viscose. J Nat Prod 62(4):601–604. doi:10.1021/
np980132u
181. Schinella GR, Tournier HA, Prieto JM et al (2002) Antioxidant activity of anti-inflammatory
plant extracts. Life Sci 70(9):1023–1033
182. León I, Mirón G, Alonso D (2006) Characterization of pentasaccharide glycosides from the
roots of Ipomoea arborescens. J Nat Prod 69(6):896–902. doi:10.1021/np0600604
183. Philpott M, Gould KS, Lim C et al (2004) In situ and in vitro antioxidant activity of sweetpo-
tato anthocyanins. J Agric Food Chem 52(6):1511–1513. doi:10.1021/jf034593j
184. Park KH, Kim JR, Lee JS, Lee H, Cho KH (2010) Ethanol and water extract of purple sweet
potato exhibits anti-atherosclerotic activity and inhibits protein glycation. J Med Food
13(1):91–98
185. Hou W-C, Chen Y-C, Chen H-J et al (2001) Antioxidant activities of trypsin inhibitor, a 33
KDa root storage protein of sweet potato (Ipomoea batatas (L.) Lam cv. Tainong 57). J Agric
Food Chem 49(6):2978–2981
186. Pereda-Miranda R, Escalante-Sánchez E, Escobedo-Martínez C (2005) Characterization of
lipophilic pentasaccharides from beach morning glory (Ipomoea pes-caprae). J Nat Prod
68(2):226–230. doi:10.1021/np0496340
187. Umamaheshwari G, Ramanathan T, Shanmugapriya R (2012) Antioxidant and radical scavenging effect of Ipomoea pes-caprae Linn. R.BR. Int J Pharm Tech Res 4(2):848–851
188. Silva DHS, Zhang Y, Santos LA et al (2005) Lipoperoxidation and cyclooxygenases 1 and 2
inhibitory compounds from Iryanthera juruensis. J Agric Food Chem 55(7):2569–2574.
doi:10.1021/jf063451x
189. Silva DH, Pereira FC, Zanoni MV, Yoshida M (2001) Lipophyllic antioxidants from
Iryanthera juruensis fruits. Phytochemistry 57(3):437–442
190. Sheng MD, López A, Hillhouse BJ et al (2002) Bioactive constituents from Iryanthera megis-
tophylla. J Nat Prod 65(10):1412–1416. doi:10.1021/np020169l
191. Ming DS, López A, Hillhouse BJ et al (2002) Bioactive constituents from Iryanthera megis-
tophylla. J Nat Prod 65(10):1412–1416

192. Jiang B, Yang H, Li M-L et al (2002) Diterpenoids from Isodon adenantha. J Nat Prod 65(8):1111–1116. doi:10.1021/np020084k

193. Wang Y-H, Chen Y-Z, Kim D-S et al (1997) Two new ent-Kauranoids from Isodon excise. J Nat Prod 60(11):1161–1162. doi:10.1021/np970155t

194. Hou A-J, Li M-L, Jiang B et al (2000) New 7,20: 14,20-diepoxy ent-kauranoids from Isodon xerophilus. J Nat Prod 63(5):599–601. doi:10.1021/np9903705

195. Hou AJ, Li ML, Jiang B, Lin ZW, Ji SY, Zhou YP, Sun HD (2000) New 7,20:14,20-diepoxy ent-kauranoids from Isodon xerophilus. J Nat Prod 63(5):599–601

196. Zhang S, Zhao M, Bai L et al (2006) Bioactive guaianolides from siyekucai (Ixeris chinensis). J Nat Prod 69(10):1425–1428. doi:10.1021/np068015j

197. Zhang S, Zhao M, Bai L et al (2006) Bioactive guaianolides from siyekucai (Ixeris chinensis). J Nat Prod 69(10):1425–1428

198. Day S-H, Chiu N-Y, Tsao L-T et al (2000) New lignan glycosides with potent antiinflammatory effect, isolated from Justicia ciliate. J Nat Prod 63(11):1560–1562. doi:10.1021/np000191j

199. Day SH, Chiu NY, Won SJ, Lin CN (1999) Cytotoxic lignans of Justicia ciliata. J Nat Prod 62(7):1056–1058

200. CorrêaGeone M, Alcântara Antônio F. de C (2012) Chemical constituents and biological activities of species of Justicia – a review, Rev. bras. Farmacogn 22(1). doi:10.1590/S0102-695X2011005000196

201. Navarro E, Alonso SJ, Trujillo J et al (2001) General behavior, toxicity, and cytotoxic activity of elenoside, a lignan from Justicia hyssopifolia. J Nat Prod 64(1):134–135. doi:10.1021/np9904861

202. Zhao Y, Yue J M, He Y N et al (1997) Eleven new eudesmane derivatives from Laggera pterodonta. J Nat Prod 60(6):545–549. doi:10.1021/np960456n

203. Liu YB, Jia W, Yao Z, Pan Q, Takaishi Y, Duan HQ (2007) Two eudesmane sesquiterpenes from Laggera pterodonta. J Asian Nat Prod Res 9(3–5):233–237

204. Cioffi G, Sanogo R, Vassallo A et al (2006) Pregnane glycosides from Leptadenia pyrotechnica. J Nat Prod 69(4):625–635. doi:10.1021/np050493r

205. Juliani HR, Wang M, Moharram H et.al (2006) Intraspecific variation in quality control parameters, polyphenol profile, and antioxidant activity in wild populations of Lippia multiflora from Ghana. In: Herbs: challenges in chemistry and biology, ACS symposium series 925, pp 126–142. doi:10.1021/bk-2006-0925.ch010

206. Hsu P-C, Huang Y-T, Tsai M-L et al (2004) Induction of apoptosis by shikonin through coordinative modulation of the Bcl-2 family, p27, and p53, release of cytochrome c, and sequential activation of caspases in human colorectal carcinoma cells. J Agric Food Chem 52(20):6330–6337. doi:10.1021/jf0495993

207. Muhammad I, Li X-C, Jacob MR et al (2003) Antimicrobial and antiparasitic (+)-trans-Hexahydrodibenzopyrans and analogues from Machaerium multiflorum. J Nat Prod 66(6):804–809. doi:10.1021/np030045o

208. Sindambiwe JB, Calomme M, Geerts S et al (1998) Evaluation of biological activities of triterpenoid saponins from Maesa lanceolata. J Nat Prod 61(5):585–590. doi:10.1021/np9705165

209. Foubert K, Breynaert A, Theunis M et al (2012) Evaluation of the anti-angiogenic activity of saponins from Maesa lanceolata by different assays. Nat Prod Commun 7(9):1149–1154

210. Huang P-L, Wang L-W, Lin C-N (1999) New triterpenoids of Mallotus repandus. J Nat Prod 62(6):891–892. doi:10.1021/np980374u

211. Lin JM, Lin CC, Chen MF, Ujiie T, Takada A (1995) Scavenging effects of Mallotus repandus on active oxygen species. J Ethnopharmacol 46(3):175–181

212. Somyote S, Jiraporn T, Somchai P et al (2001) D:A Friedo-oleanane Lactones from the Stems of Mallotus repandus. J Nat Prod 64(5):569–571. doi:10.1021/np0005560

213. Tazawa K, Ohkami H, Yamashita I, Ohnishi Y, Saito T, Okamoto M, Masuyama K, Yamazaki K, Takemori S, Saito M, Arai H (1998) Anticarcinogenic and/or Antimetastatic action of apple pectin in Experimental Rat Colon Carcinogenesis and on Hepatic Metastasis Rat

Model, functional foods for disease prevention I. ACS Symp Ser 701:96–103. doi:10.1021/bk-1998-0701.ch009

214. Seeram NP, Cichewicz RH, Chandra A et al (2003) Cyclooxygenase inhibitory and antioxidant compounds from crabapple fruits. J Agric Food Chem 51(7):1948–1951. doi:10.1021/jf025993u

215. Nakagawa H, Takaishi Y, Fujimoto Y et al (2004) Chemical constituents from the Colombian medicinal plant Maytenus laevis. J Nat Prod 67(11):1919–1924. doi:10.1021/np040006s

216. Kim H-J, Chen F, Wu C et al (2004) Evaluation of antioxidant activity of Australian tea tree (Melaleuca alternifolia) oil and its components. J Agric Food Chem 52(10):2849–2854. doi:10.1021/jf035377d

217. Hung C-Y, Yen G-C (2002) Antioxidant activity of phenolic compounds isolated from Mesona procumbens Hemsl. J Agric Food Chem 50(10):2993–2997. doi:10.1021/jf011454y

218. Lai LS, Chou ST, Chao WW (2001) Studies on the antioxidative activities of Hsian-tsao (Mesona procumbens Hemsl) leaf gum. J Agric Food Chem 49(2):963–968

219. Jin W, Zjawiony JK (2006) 5-Alkylresorcinols from Merulius incarnates. J Nat Prod 69(4):704–706. doi:10.1021/np050520d

220. Zjawiony JK, Jin W, Vilgalys R (2005) Merulius incarnates Schwein., a rare mushroom with highly selective antimicrobial activity. Int J Med Mushrooms 7:365–366

221. Chen J-J, Chou T-H, Peng C-F et al (2007) Antitubercular Dihydroagarofuranoid Sesquiterpenes from the roots of Microtropis fokienensis. J Nat Prod 70(2):202–205. doi:10.1021/np060500r

222. Chen I-H, Lu M-C, Du Y-C et al (2009) Cytotoxic triterpenoids from the stems of Microtropis japonica. J Nat Prod 72(7):1231–1236. doi:10.1021/np800694b

223. Catalán CAN, Cuenca M d R, Hernández LR et al (2003) cis, cis-Germacranolides and Melampolides from Mikania thapsoides. J Nat Prod 66(7):949–953. doi:10.1021/np030055p

224. Laurella LC, Frank FM, Sarquiz A, Alonso MR, Giberti G, Cavallaro L, Catalán CA, Cazorla SI, Malchiodi E, Martino VS, Sülsen VP (2012) In vitro evaluation of antiprotozoal and antiviral activities of extracts from Argentinean mikania species. Sci World J 2012:121253. doi:10.1100/2012/121253

225. Akihisa T, Tokuda H, Yasukawa K et al (2005) Azaphilones, furanoisophthalides, and amino acids from the extracts of Monascus pilosus-fermented rice (Red-mold rice) and their chemopreventive effects. J Agric Food Chem 53(3):562–565. doi:10.1021/jf040199p

226. Chen CC, Chyau CC, Liao CC, Hu TJ, Kuo CF (2010) Enhanced anti-inflammatory activities of Monascus pilosus fermented products by addition of ginger to the medium. J Agric Food Chem 58(22):12006–12013

227. Müller S, Murillo R, Castro V et al (2004) Sesquiterpene lactones from Montanoa hibiscifolia that inhibit the transcription factor NF-κB. J Nat Prod 67(4):622–630. doi:10.1021/np034072q

228. Bagnarello G, Hilje L, Bagnarello V, Cartín V, Calvo M (2009) Phagodeterrent activity of the plants Tithonia diversifolia and Montanoa hibiscifolia (Asteraceae) on adults of the pest insect Bemisia tabaci (Homoptera: Aleyrodidae). Rev Biol Trop 57(4):1201–1215

229. Yu H, Li S, Huang M-T et.al (2008) Antiinflammatory constituents in Noni (Morinda citrifolia) fruits, ACS symposium series, 987. doi:10.1021/bk-2008-0987.ch012

230. Gilani AH, Mandukhail S-u-R, Iqbal J et al (2010) Antispasmodic and vasodilator activities of Morinda citrifolia root extract are mediated through blockade of voltage dependent calcium channels. BMC Complement Altern Med 10:2. doi:10.1186/1472-6882-10-2

231. Bennett RN, Mellon FA, Foidl N et al (2003) Profiling glucosinolates and phenolics in vegetative and reproductive tissues of the multi-purpose trees Moringa oleifera L. (Horseradish tree) and Moringa stenopetala L. J Agric Food Chem 51(12):3546–3553. doi:10.1021/jf0211480

232. Walter A, Samuel W, Peter A et al (2011) Antibacterial activity of Moringa oleifera and Moringa stenopetala methanol and n-hexane seed extracts on bacteria implicated in water borne diseases. Afr J Microbiol Res 5(2):153–157. doi:10.5897/AJMR10.457

233. Pawlowska AM, Oleszek W, Braca A (2008) Quali-quantitative analyses of Flavonoids of Morus nigra L. and Morus alba L. (Moraceae) fruits. J Agric Food Chem 56(9):3377–3380. doi:10.1021/jf703709r

234. Wang L, Yang Y, Liu C, Chen RY (2010) Three new compounds from Morus nigra L. J Asian Nat Prod Res 12(6):431–437
235. Ramsewak RS, Nair MG, Strasburg GM et al (1999) Biologically active carbazole alkaloids from Murraya koenigii. J Agric Food Chem 47(2):444–447. doi:10.1021/jf9805808
236. Gupta BK, Tailang M, Lokhande AK et al (2011) Antimicrobial activity of ethanolic extracts of Murraya Koenigii by disc diffusion and broth dilution method. J Pharm Res 4(4):1023
237. Mathur A (2011) Antiinflammatory activity of leaves extracts of murraya koenigii L. Int J Pharma Bio Sci 2(1):541
238. Sawadjoon S, Kittakoop P, Kirtikara K et al (2002) Selective COX-2 inhibitors and antifungal agents from Myristica cinnamomea. J Org Chem 67(16):5470–5475. doi:10.1021/jo020045d
239. Chong YM, Yin WF, Ho CY et al (2011) Malabaricone C from Myristica cinnamomea exhibits anti-quorum sensing activity. J Nat Prod 74(10):2261–2264
240. Charan RD, Munro MHG, O'Keefe BR et al (2000) Isolation and characterization of Myrianthus holstii Lectin a Potent HIV-1 inhibitory protein from the plant Myrianthus holstii. J Nat Prod 63(8):1170–1174. doi:10.1021/np000039h
241. Bi Y, Yang G, Li H et al (2006) Characterization of the chemical composition of lotus plumule oil. J Agric Food Chem 54(20):7672–7677. doi:10.1021/jf0607011
242. Markin D, Duek L, Berdicevsky I (2003) In vitro antimicrobial activity of olive leaves. Mycoses 46(3–4):132–136
243. Tabera J, Guinda Á, Rodríguez AR et al (2004) Countercurrent supercritical fluid extraction and fractionation of high-added-value compounds from a hexane extract of olive leaves. J Agric Food Chem 52(15):4774–4779. doi:10.1021/jf049881+
244. Bayçın D, Altıok E, Ülkü S et al (2007) Adsorption of olive leaf (Olea europaea L.) antioxidants on silk fibroin. J Agric Food Chem 55(4):1227–1236. doi:10.1021/jf062829o
245. Kırmızıgül S, Gören N, Yang S-W et al (1997) Spinonin, a novel glycoside from ononis spinosa subsp. Leiosperma. J Nat Prod 60(4):378–381. doi:10.1021/np9605652
246. Altuner EM, Ceter T, Işlek C (2010) Investigation of antifungal activity of Ononis spinosa L. ash used for the therapy of skin infections as folk remedies. Mikrobiyol Bul 44(4):633–639
247. Galati EM, Mondello MR, Giuffrida D et al (2003) Chemical characterization and biological effects of Sicilian Opuntia ficus indica (L.) mill. Fruit juice: antioxidant and antiulcerogenic activity. J Agric Food Chem 51(17):4903–4908. doi:10.1021/jf030123d
248. Demirci F, Paper DH, Franz G et al (2004) Investigation of the Origanum onites L. essential oil using the Chorioallantoic Membrane (CAM) Assay. J Agric Food Chem 52(2):251–254. doi:10.1021/jf034850k
249. Sarac N, Ugur A (2008) Antimicrobial activities of the essential oils of Origanum onites L., Origanum vulgare L. subspecies hirtum (Link) Ietswaart, Satureja thymbra L., and Thymus cilicicus Boiss. & Bal. growing wild in Turkey. J Med Food 11(3):568–573
250. Tang Y, Yu B, Hu J et al (2002) Three new homoisoflavanone glycosides from the bulbs of Ornithogalum caudatum. J Nat Prod 65(2):218–220. doi:10.1021/np010466a
251. Chen R, Meng F, Liu Z et al (2010) Antitumor activities of different fractions of polysaccharide purified from Ornithogalum caudatum Ait. Carbohydr Polym 80(3):845–851. doi:10.1016/j.carbpol. 12.042
252. Awale S, Tezuka Y, Banskota AH et al (2003) Nitric oxide inhibitory isopimarane-type diterpenes from Orthosiphon stamineus of Indonesia. J Nat Prod 66(2):255–258. doi:10.1021/np020455x
253. Abdelwahab SI, Mohan S, Elhassan MM et.al (2011) Antiapoptotic and Antioxidant properties of Orthosiphon stamineus Benth (Cat's Whiskers): intervention in the Bcl-2-Mediated Apoptotic Pathway. Evid Based Complement Alternat Med. 156765. doi:10.1155/2011/156765
254. Lin H-C, Ding H-Y, Wu Y-C (1998) Two novel compounds from Paeonia suffruticosa. J Nat Prod 61(3):343–346. doi:10.1021/np9704258
255. Tak JH, Kim HK, Lee SH et al (2006) Acaricidal activities of paeonol and benzoic acid from Paeonia suffruticosa root bark and monoterpenoids against Tyrophagus putrescentiae (Acari: Acaridae). Pest Manag Sci 62(6):551–557
256. Noh IC, Cho WD, Sandesh S et al (2012) Anti-inflammatory and immunosuppressive activity of mixture of Trachelospermum asiaticum and Paeonia suffruticosa extracts (novel herbal formula SI000902). J Med Plant Res 6(25):4247–4254. doi:10.5897/JMPR12.509

257. Kang KS, Yokozawa T, Kim HY et al (2006) Study on the nitric oxide scavenging effects of ginseng and its compounds. J Agric Food Chem 54(7):2558–2562. doi:10.1021/jf0529520

258. Ramesh T, Kim SW, Sung JH, Hwang SY, Sohn SH, Yoo SK, Kim SK (2011) Effect of fermented Panax ginseng extract (GINST) on oxidative stress and antioxidant activities in major organs of aged rats. Exp Gerontol 47(1):77–84

259. Zou K, Zhu S, Tohda C et al (2002) Dammarane-type triterpene saponins from Panax japonicas. J Nat Prod 65(3):346–351. doi:10.1021/np010354j

260. Chan H-H, Hwang T-L, Sun H-D et al (2011) Bioactive constituents from the roots of Panax japonicus var. MAJOR and development of a LC-MS/MS method for distinguishing between natural and artifactual compounds. J Nat Prod 74(4):796–802. doi:10.1021/np100851s

261. Peng Y, Ye J, Kong J (2005) Determination of phenolic compounds in Perilla frutescens L. by capillary electrophoresis with electrochemical detection. J Agric Food Chem 53(21):8141–8147. doi:10.1021/jf051360e

262. Meng L, Lozano YF, Gaydou EM et al (2009) Antioxidant activities of polyphenols extracted from Perilla frutescens varieties. Molecules 14:133–140. doi:10.3390/molecules14010133

263. Saita E, Kishimoto Y, Tani M et al (2012) Antioxidant activities of Perilla frutescens against low-density lipoprotein oxidation in vitro and in human subjects. J Oleo Sci 61(3):113–120

264. Kang R, Helms R, Stout MJ et al (1992) Antimicrobial activity of the volatile constituents of Perilla frutescens and its synergistic effects with polygodial. J Agric Food Chem 40(11):2328–2330. doi:10.1021/jf00023a054

265. Vogl S, Zehl M, Picker P et al (2011) Identification and quantification of coumarins in Peucedanum ostruthium (L.) Koch by HPLC-DAD and HPLC-DAD-MS. J Agric Food Chem 59(9):4371–4377. doi:10.1021/jf104772x

266. Hiermann A, Schantl D (1998) Antiphlogistic and antipyretic activity of Peucedanum ostruthium. Planta Med 64(5):400–403

267. Ahmad S, Malik A, Afza N et al (1999) New withanolide glycoside from Physalis peruviana. J Nat Prod 62(3):493–494. doi:10.1021/np980228o

268. Wu SJ, Ng LT, Huang YM et al (2005) Antioxidant activities of Physalis peruviana. Biol Pharm Bull 28(6):963–966

269. Zhang Y-J, Tanaka T, Iwamoto Y, Yang C-R, Kouno I (2001) Novel sesquiterpenoids from the roots of Phyllanthus emblica. J Nat Prod 64(7):870–873. doi:10.1021/np010059z

270. Bandyopadhyay SK, Pakrashi SC, Pakrashi A (2000) The role of antioxidant activity of Phyllanthus emblica fruits on prevention from indomethacin induced gastric ulcer. J Ethnopharmacol 70(2):171–176

271. Jia Q, Hong M-F, Minter PD (1992) A novel irid from Picrorhiza kurro. J Nat Prod 62(6):901–903. doi:10.1021/np980493

272. Rajkumar V, Guha G, Kumar RA (2011) Antioxidant and anti-neoplastic activities of Picrorhiza kurroa extracts. Food Chem Toxicol 49(2):363–369. doi:10.1016/j.fct.2010.11.009

273. Smit HF, Kroes BH, van den Berg AJ, van der Wal D, van den Worm E, Beukelman CJ, van Dijk H, Labadie RP (2000) Immunomodulatory and anti-inflammatory activity of Picrorhiza scrophulariiflora. J Ethnopharmacol 73(1–2):101–109

274. Smit HF, van den Berg AJJ, Kroes BH (2000) Inhibition of T-lymphocyte proliferation by cucurbitacins from Picrorhiza scrophulariaeflora. J Nat Prod 63(9):1300–1302. doi:10.1021/np990215q

275. Grassmann J, Hippeli S, Vollmann R et al (2003) Antioxidative properties of the essential oil from Pinus mugo. J Agric Food Chem 51(26):7576–7582. doi:10.1021/jf030496e

276. Wei K, Li W, Koike K (2005) Nigramides A-S, dimeric amide alkaloids from the roots of Piper nigrum. J Org Chem 70(4):1164–1176. doi:10.1021/jo040272a

277. Park IK, Lee SG, Shin SC (2002) Larvicidal activity of isobutylamides identified in Piper nigrum fruits against three mosquito species. J Agric Food Chem 50(7):1866–1870

278. Alma MH, Nitz S, Kollmannsberger H (2004) Chemical composition and antimicrobial activity of the essential oils from the gum of Turkish pistachio (Pistacia vera L.). J Agric Food Chem 52(12):3911–3914. doi:10.1021/jf040014e

279. Chiang LC, Chiang W, Chang MY et al (2002) Antiviral activity of Plantago major extracts and related compounds in vitro. Antivir Res 55(1):53–62

280. Vastano C, Rafi MM, DiPaola RS et al (2001) Bioactive homoisoflavones from Vietnamese Coriander or Pak Pai (Polygonatum odoratum). In: Quality management of nutraceuticals, ACS symposium series, 803, pp 269–280. doi:10.1021/bk-2002-0803.ch019

281. Wang D, Zeng L, Li D et al (2013) Antioxidant activities of different extracts and homo iso-flavanones isolated from the Polygonatum odoratum. Nat Prod Res 27(12):1111–1114. doi:1 0.1080/14786419.2012.701212

282. Sang S, Lapsley K, Rosen RT et al (2002) New prenylated benzoic acid and other constituents from almond hulls (Prunus amygdalus Batsch). J Agric Food Chem 50(3):607–609. doi:10.1021/jf0110194

283. Wang H, Nair MG, Strasburg GM et al (2003) Antioxidant and antiinflammatory activities of anthocyanins and their aglycon, cyanidin, from tart cherries. J Nat Prod 62(2):294–296. doi:10.1021/np980501m

284. Takeoka GR, Dao LT (2003) Antioxidant constituents of almond [Prunus dulcis (Mill.) D.A. Webb] hulls. J Agric Food Chem 51(2):496–501. doi:10.1021/jf020660i

285. Hamauzu Y, Kume C, Yasui H et al (2007) Reddish coloration of Chinese quince (Pseudocydonia sinensis) procyanidins during heat treatment and effect on antioxidant and antiinfluenza viral activities. J Agric Food Chem 55(4):1221–1226. doi:10.1021/jf061836+

286. Demuner AJ, Barbosa LCdA, Howarth OW (1996) Structure and plant growth regulatory activity of new diterpenes from Pterodon polygalaeflorus. J Nat Prod 59(8):770–772. doi:10.1021/np960140f

287. De Omena MC, Bento ES, De Paula JE et al (2006) Larvicidal diterpenes from Pterodon polygalaeflorus. Vector Borne Zoonotic Dis 6(2):216–222. doi:10.1089/vbz.2006.6.216

288. Lee H-S (2002) Tyrosinase inhibitors of Pulsatilla cernua root-derived materials. J Agric Food Chem 50(6):1400–1403. doi:10.1021/jf011230f

289. Lee HS, Beon MS, Kim MK (2001) Selective growth inhibitor toward human intestinal bacteria derived from Pulsatilla cernua root. J Agric Food Chem 49(10):4656–4661

290. Dell'Agli M, Galli GV, Corbett Y et al (2009) Antiplasmodial activity of Punica granatum L. fruit rind. J Ethnopharmacol 125(2):279–285

291. Ahmed S, Wang N, Hafeez BB et al (2005) Punica granatum L. extract inhibits IL-1β–induced expression of matrix metalloproteinases by inhibiting the activation of MAP kinases and NF-κB in human chondrocytes in vitro. J Nutr 135(9):2096–2102

292. Krenn L, Presser A, Pradhan R et al (2003) Sulfemodin 8-O-β-d-glucoside, a new sulfated anthraquinone glycoside, and antioxidant phenolic compounds from Rheum emodi. J Nat Prod 66(8):1107–1109. doi:10.1021/np0301442

293. Iwata N, Wang N, Yao X et al (2004) Structures and histamine release inhibitory effects of prenylated orcinol derivatives from Rhododendron dauricum. J Nat Prod 67(7):1106–1109. doi:10.1021/np0303916

294. Goffman FD, Galletti S (2001) Gamma-linolenic acid and tocopherol contents in the seed oil of 47 accessions from several ribes species. J Agric Food Chem 49(1):349–354. doi:10.1021/jf0006729

295. Knox YM, Suzutani T, Yosida I et al (2003) Anti-influenza virus activity of crude extract of Ribes nigrum L. Phytother Res 17(2):120–122

296. Larsen E, Kharazmi A, Christensen LP et al (2003) An antiinflammatory galactolipid from rose hip (Rosa canina) that inhibits chemotaxis of human peripheral blood neutrophils in vitro. J Nat Prod 66(7):994–995. doi:10.1021/np0300636

297. Kilicgun H, Altiner D (2010) Correlation between antioxidant effect mechanisms and poly-phenol content of Rosa canina. Pharmacognosy Res 6(23):238–241. doi:10.4103/0973-1296.66943

298. Lattanzio F, Greco E, Carretta D et al (2011) In vivo anti-inflammatory effect of Rosa canina L. extract. J Ethnapharmacol 137(1):880–885

299. Kirkeskov B, Christensen R, Bügel S et al (2011) The effects of rose hip (Rosa canina) on plasma antioxidative activity and C-reactive protein in patients with rheumatoid arthritis and normal controls: a prospective cohort study. Phytomedicine 18(11):953–958

300. Altinier LG, Sosa S, Aquino RP et al (2007) Characterization of topical antiinflammatory compounds in Rosmarinus officinalis. J Agric Food Chem 55(5):1718–1723. doi:10.1021/jf062610

301. Ventura-Martínez R, Rivero-Osorno O, Gómez C et al (2011) Spasmolytic activity of Rosmarinus officinalis L. involves calcium channels in the guinea pig ileum. J Ethnopharmacol 137(3):1528–1532

302. Naito Y, Oka S, Yoshikawa T (2003) Inflammatory response in the Pathogenesis of Atherosclerosis and its prevention by Rosmarinic Acid, a functional ingredient of Rosemary. In: Food factors in health promotion and disease prevention, ACS symposium series, 851, pp 208–221. doi:10.1021/bk-2003-0851.ch018

303. Wang B-G, Zhu W-M, Li X-M et al (2000) Rubupungenosides A and B, two novel triterpenoid saponin dimers from the aerial parts of Rubus pungens. J Nat Prod 63(6):851–854. doi:10.1021/np990473n

304. Wada L, Ou B (2002) Antioxidant activity and phenolic content of Oregon Caneberries. J Agric Food Chem 50(12):3495–3500. doi:10.1021/jf0114051

305. Bushman BS, Phillips B, Isbell T et al (2004) Chemical composition of Caneberry (Rubus spp.) seeds and oils and their antioxidant potential. J Agric Food Chem 52(26):7982–7987. doi:10.1021/jf049149a

306. Longo L, Vasapollo G (2005) Determination of anthocyanins in Ruscus aculeatus L. Berries. J Agric Food Chem 53(2):475–479. doi:10.1021/jf0487250

307. Facino RM, Carini M, Stefani R et al (2006) Anti-Elastase and Anti-Hyaluronidase activities of saponins and sapogenins from Hedera helix, Aesculus hippocastanum, and Ruscus aculeatus: factors contributing to their efficacy in the treatment of venous insufficiency. Arch Pharm 328(10):720–724. doi:10.1002/ardp.19953281006

308. Chen X-H, Xia L-X, Hong-Bo H-B et al (2010) Chemical composition and antioxidant activities of Russula griseocarnosa sp. nov. J Agric Food Chem 58(11):6966–6971. doi:10.1021/jf1011775

309. Morikawa T, Kishi A, Pongpiriyadacha Y et al (2003) Structures of new friedelane-type triterpenes and eudesmane-type sesquiterpene and aldose reductase inhibitors from Salacia chinensis. J Nat Prod 66(9):1191–1196. doi:10.1021/np0301543

310. Sikarwar MS, Patil MB (2012) Antihyperlipidemic activity of Salacia chinensis root extracts in triton-induced and atherogenic diet-induced hyperlipidemic rats. Indian J Pharm 44(1):88–92

311. Fraga BM, Daz CE, Guadao A et al (2005) Diterpenes from Salvia broussonetii transformed roots and their insecticidal activity. J Agric Food Chem 53(13):5200–5206. doi:10.1021/jf058045c

312. Santos-Gomes PC, Fernandes-Ferreira M (2003) Essential oils produced by in vitro shoots of sage (Salvia officinalis L.). J Agric Food Chem 51(8):2260–2266. doi:10.1021/jf020945v

313. Bouajaj S, Benyamna A, Bouamama H et al (2012) Antibacterial, allelopathic and antioxidant activities of essential oil of Salvia officinalis L. growing wild in the Atlas Mountains of Morocco. Nat Prod Res 27(18):1673–1676. doi:10.1080/14786419.2012.751600

314. Don M-J, Shen C-C, Lin Y-L et al (2005) Nitrogen-containing compounds from Salvia miltiorrhiza. J Nat Prod 68(7):1066–1070. doi:10.1021/np0500934

315. Zhao G-R, Xiang Z-J, Ye T-X et al (2006) Antioxidant activities of Salvia miltiorrhiza and Panax notoginseng. Food Chem 99(4):767–774

316. Pan Z-H, Wang Y-Y, Li M-M et al (2010) Terpenoids from Salvia trijuga. J Nat Prod 73(6):1146–1150. doi:10.1021/np100250w

317. Fattorusso E, Santelia FU, Appendino G et al (2004) Polyoxygenated eudesmanes and trans-chrysanthemanes from the aerial parts of Santolina insularis. J Nat Prod 67(1):37–41. doi:10.1021/np0302221

318. Valenti D, De Logu A, Loy G et al (2001) Liposome-incorporated santolina insularis essential oil: preparation, characterization and in vitro antiviral activity. J Liposome Res 11(1):73–90. doi:10.1081/LPR-100103171

319. Silván AM, Abad MJ, Bermejo P et al (1996) Antiinflammatory activity of coumarins from Santolina oblongifolia. J Nat Prod 59(12):1183–1185. doi:10.1021/np960422f

320. Ogundaini A, Farah M, Perera P et al (1996) Isolation of two new antiinflammatory biflavanoids from Sarcophyte pirie. J Nat Prod 59(6):587–590. doi:10.1021/np960386k

321. Selenski C, Pettus TRR (2006) (±)-Diinsininone: made nature's way. Tetrahedron 62:5298–5307

322. Chorianopoulos N, Evergetis E, Mallouchos A et al (2006) Characterization of the essential oil volatiles of Satureja thymbra and Satureja parnassica: influence of harvesting time and antimicrobial activity. J Agric Food Chem 54(8):3139–3145. doi:10.1021/jf053183n

323. Sun C-M, Syu W-J, Don M-J et al (2003) Cytotoxic sesquiterpene lactones from the root of Saussurea lappa. J Nat Prod 66(9):1175–1180. doi:10.1021/np030147e

324. Gokhale AB, Damre AS, Kulkami KR et al (2002) Preliminary evaluation of anti-inflammatory and anti-arthritic activity of S. lappa, A. speciosa and A. aspera. Phytomedicine 9(5):433–437

325. Heo HJ, Kim D-O, Choi SJ et al (2004) Potent inhibitory effect of flavonoids in Scutellaria baicalensis on amyloid β protein-induced neurotoxicity. J Agric Food Chem 52(13):4128–4132. doi:10.1021/jf049953x

326. Joshee N, Tascan A, Medina-Bolivar F et al (2013) Scutellaria: biotechnology, phytochemistry and its potential as a commercial medicinal crop. In: Chandra S et al (eds) Micropropagation and improvement. Springer, Heidelberg, pp 69–99

327. Ye F, Xui L, Yi J, Zhang W, Zhang DY (2002) Anticancer activity of Scutellaria baicalensis and its potential mechanism. J Altern Complement Med 8(5):567–572

328. Šentjurc M, Nemec M, Connor HD et al (2003) Antioxidant activity of Sempervivum tectorum and its components. J Agric Food Chem 51(9):2766–2771. doi:10.1021/jf026029z

329. Panzella L, Eidenberger T, Napolitano A et al (2012) Black sesame pigment: DPPH assay-guided purification, antioxidant/antinitrosating properties and identification of a degradative structural marker. J Agric Food Chem 60(36):8895–8901. doi:10.1021/jf2053096

330. Sautour M, Miyamoto T, Lacaille-Dubois M-A (2005) Steroidal saponins from Smilax medica and their antifungal activity. J Nat Prod 68(10):1489–1493. doi:10.1021/np058060b

331. Howard LR, Pandjaitan N, Morelock T, Gil MI (2002) Antioxidant capacity and phenolic content of spinach as affected by genetics and growing season. J Agric Food Chem 50(21):5891–5896. doi:10.1021/jf020507o

332. Kumar US, Tiwari AK, Reddy SV et al (2005) Free-radical-scavenging and xanthine oxidase inhibitory constituents from Stereospermum personatum. J Nat Prod 68(11):1615–1621. doi:10.1021/np058036y

333. Barthomeuf CM, Debiton E, Barbakadze VV et al (2001) Evaluation of the dietetic and therapeutic potential of a high molecular weight hydroxycinnamate-derived polymer from symphytum asperum lepech. Regarding its antioxidant, antilipoperoxidan, antiinflammatory, and cytotoxic properties. J Agric Food Chem 49(8):3942–3946. doi:10.1021/jf010189d

334. Barbakadze V, Kemertelidze E, Targamadze I et al (2005) Poly [3-(3,4-dihydroxyphenyl) glyceric acid], a new biologically active polymer from symphytum asperum lepech. and S. Caucasicum Bieb. (boraginaceae). Molecules 10:1135–1144

335. Park B-S, Kim J-R, Lee S-E, Kim KS et al (2005) Selective growth-inhibiting effects of compounds identified in Tabebuia impetiginosa inner bark on human intestinal bacteria. J Agric Food Chem 53(4):1152–1157. doi:10.1021/jf0486038

336. Park BS, Lee HK, Lee SE, Piao XL, Takeoka GR, Wong RY, Ahn YJ, Kim JH (2006) Antibacterial activity of Tabebuia impetiginosa Martius ex DC (Taheebo) against Helicobacter pylori. J Ethanophramacol 105(1–2):255–262

337. Xiang W, Li R-T, Mao Y-L et al (2005) Four new prenylated isoflavonoids in Tadehagi triquetrum. J Agric Food Chem 53(2):267–271. doi:10.1021/jf0483117

338. Céspedes CL, Avila JG, Martínez A et al (2006) Antifungal and antibacterial activities of Mexican Tarragon (Tagetes lucida). J Agric Food Chem 54(10):3521–3527. doi:10.1021/jf053071w

339. Martinez J, Silván AM, Abad MJ et al (1997) Isolation of two flavonoids from Tanacetum microphyllum as PMA-induced ear edema inhibitors. J Nat Prod 60(2):142–144. doi:10.1021/np960163u

340. Abad MJ, Bermejo P, Villar A, Valverde S (1993) Anti-inflammatory activity of two flavonoids from Tanacetum microphyllum. J Nat Prod 56(7):1164–1167

341. Abad MJ, Bermejo PA (1991) Antiinflammatory and anti-ulcerogenic activities of the organic extracts of Tanacetum microphyllum DC in rats. Villar Phytother Res 5(4):179–181. doi:10.1002/ptr.2650050409

342. Ndubuisil MKI, Kwok BHB, Vervoort J et al (2002) Characterization of a novel mammalian phosphatase having sequence similarity to schizosaccharomyces pombe PHO2 and Saccharomyces cerevisiae PHO13. Biochemistry 41(24):7841–7848. doi:10.1021/bi0255064

343. Majdi M, Liu Q, Karimzadeh G, Malboobi MA, Beekwilder J, Cankar K, Vos R, Todorović S, Simonović A, Bouwmeester H (2011) Biosynthesis and localization of parthenolide in glandular trichomes of feverfew (Tanacetum parthenium L. Schulz Bip.). Phytochemistry 72(14–15):1739–1750

344. Lesiak K, Koprowska K, Zalesna I, Nejc D, Düchler M, Czyz M (2010) Parthenolide, a sesquiterpene lactone from the medical herb feverfew, shows anticancer activity against human melanoma cells in vitro. Melanoma Res 20(1):21–34

345. Barrero AF, Quílez del Moral JF, Lucas R et al (2003) Diterpenoids from Tetraclinis articulata that inhibit various human leukocyte functions. J Nat Prod 66(6):844–850. doi:10.1021/np0204949

346. Djouahri A, Boudarene L, Meklati BY (2013) Effect of extraction method on chemical composition, antioxidant and anti-inflammatory activities of essential oil from the leaves of Algerian Tetraclinis articulata (Vahl) masters. Ind Crop Prod 44:32–36

347. Ghosal S, Vishwakarma RA (1997) Tinocordiside, a new rearranged cadinane sesquiterpene glycoside from Tinospora cordifolia. J Nat Prod 60(8):839–841. doi:10.1021/np970169z

348. Stanely P, Prince M, Menon VP (2000) Hypoglycaemic and other related actions of Tinospora cordifolia roots in alloxan-induced diabetic rats. J Ethnopharmacol 70:9–15. doi:10.1016/S0378-8741(98)00164-0

349. Desai VR, Kamat JP, Sainis KB (2002) An immunomodulator from Tinospora cordifolia with antioxidant activity in cell-free systems. Proc Indian Acad Sci 114(6):713–719

350. Lin Y-L, Tsai Y-L, Kuo Y-H et al (1999) Phenolic compounds from Tournefortia sarmentosa. J Nat Prod 62(11):1500–1503. doi:10.1021/np9901332

351. Lin YL, Chang YY, Kuo YH, Shiao MS (2002) Anti-lipid-peroxidative principles from Tournefortia sarmentosa. J Nat Prod 65(5):745–747

352. Miyazawa M, Okuno Y, Imanishi K (2005) Suppression of the SOS-inducing activity of mutagenic heterocyclic amine, Trp-P-1, by triterpenoid from Uncaria sinensis in the Salmonella typhimurium TA1535/pSK1002 Umu test. J Agric Food Chem 53(6):2312–2315. doi:10.1021/jf035430y

353. Park SH, Kim JH, Park SJ, Bae SS, Choi YW, Hong JW, Choi BT, Shin HK (2011) Protective effect of hexane extracts of Uncaria sinensis against photothrombotic ischemic injury in mice. J Ethnopharmacol 138(3):774–779. doi:10.1016/j.jep.2011.10.026

354. Neto CC (2011) Ursolic acid and other pentacyclic triterpenoids: Anticancer activities and occurrence in berries. In: Stoner GD, Seeram NP (eds) Berries and cancer prevention. Springer Science+Business Media, LLC, New York. doi:10.1007/978-1-4419-7554-6

355. Kitajima M, Hashimoto K-i, Yokoya M et al (2003) Two new nor-triterpene glycosides from Peruvian Uña de Gato (Uncaria tomentosa). J Nat Prod 66(2):320–323. doi:10.1021/np0203741

356. Pilarski R, Poczekaj-Kostrzewska M, Ciesiołka D et al (2007) Antiproliferative activity of various Uncaria tomentosa preparations on HL-60 promyelocytic leukemia cells. Pharmacol Rep 59(5):565–572

357. Ayaz FA, Hayirlioglu-Ayaz S, Gruz J et al (2005) Separation, characterization, and quantitation of phenolic acids in a little-known blueberry (Vaccinium arctostaphylos L.) fruit by HPLC-MS. J Agric Food Chem 53(21):8116–8122. doi:10.1021/jf058057y
358. Su Z (2012) Anthocyanins and flavonoids of vaccinium L. Pharm Crops 3:7–37
359. Dulebohn RV, Yi W, Srivastava A et al (2008) Effects of blueberry (Vaccinium ashei) on DNA damage, lipid peroxidation, and phase II enzyme activities in rats. J Agric Food Chem 56(24):11700–11706. doi:10.1021/jf802405y
360. Li C, Feng J, Huang WY, An XT (2013) Composition of polyphenols and antioxidant activity of rabbiteye blueberry (Vaccinium ashei) in Nanjing. J Agric Food Chem 61(3):523–531. doi:10.1021/jf3046158
361. Dao CA, Patel KD, Neto CC (2012) Phytochemicals from the fruit and foliage of Cranberry (Vaccinium macrocarpon) – potential benefits for human health. In: Emerging trends in dietary components for preventing and combating disease, ACS symposium series, 1093, pp 79–94. doi:10.1021/bk-2012-1093.ch005
362. Yan X, Murphy BT, Hammond GB, Vinson JA, Neto CC (2002) Antioxidant activities and antitumor screening of extracts from cranberry fruit (Vaccinium macrocarpon). J Agric Food Chem 50(21):5844–5849
363. Murphy BT, MacKinnon SL, Yan X et al (2003) Identification of triterpene hydroxycinnamates with in vitro antitumor activity from whole cranberry fruit (Vaccinium macrocarpon). J Agric Food Chem 51(12):3541–3545. doi:10.1021/jf034114g
364. Bao L, Yao X-S, Tsi D et al (2008) Protective effects of bilberry (Vaccinium myrtillus L.) extract on KBrO₃-induced kidney damage in mice. J Agric Food Chem 56(2):420–425. doi:10.1021/jf072640s
365. Szakie A, Pączkowski C, Koivuniemi H et al (2012) Comparison of the triterpenoid content of berries and leaves of lingonberry vaccinium vitis-idaea from Finland and Poland. J Agric Food Chem 60(19):4994–5002. doi:10.1021/jf300375b
366. Ho KY, Tsai CC, Huang JS, Chen CP, Lin TC, Lin CC (2001) Antimicrobial activity of tannin components from Vaccinium vitis-idaea L. J Pharm Pharmacol 53(2):187–191
367. Kylli P, Nohynek L, Puupponen-Pimia R et al (2011) Lingonberry (Vaccinium vitis-idaea) and European Cranberry (Vaccinium microcarpon) proanthocyanidins: isolation, identification, and bioactivities. J Agric Food Chem 59(7):3373–3384. doi:10.1021/jf104621e
368. de Andrade IL, Bezerra JNS, Lima MAA et al (2004) Chemical composition and insecticidal activity of essential oils from Vanillosmopsis pohlii Baker against Bemisia argentifolii. J Agric Food Chem 52(19):5879–5881. doi:10.1021/jf049788l
369. Zgoda-Pols JR, Freyer AJ, Killme LB, Porter JR (2002) Antimicrobial resveratrol tetramers from the stem bark of Vatica oblongifolia ssp. Oblongifolia. J Nat Prod 65(11):1554–1559. doi:10.1021/np020198w
370. Zgoda-Pols JR, Freyer AJ, Killmer LB, Porter JR (2002) Antimicrobial resveratrol tetramers from the stem bark of Vatica oblongifoliassp. oblongifolia. J Nat Prod 65(11):1554–1559
371. Hernández-Pérez M, Hernández T, Gómez-Cordovés C et al (1996) Phenolic composition of the "Mocán" (Visnea mocanera L.f.). J Agric Food Chem 44(11):3512–3515. doi:10.1021/jf9505335
372. Castillo J, Benavente-García O, Lorente J et al (2000) Antioxidant activity and radioprotective effects against chromosomal damage induced in vivo by X-rays of flavan-3-ols (Procyanidins) from grape seeds (Vitis vinifera): comparative study versus other phenolic and organic compounds. J Agric Food Chem 48(5):1738–1745. doi:10.1021/jf990665o
373. Terra X, Valls J, Vitrac X et al (2007) Grape-seed procyanidins act as antiinflammatory agents in endotoxin-stimulated RAW 264.7 macrophages by inhibiting NFkB signaling pathway. J Agric Food Chem 55(11):4357–4365. doi:10.1021/jf0633185
374. Jayaprakasha GK, Singh RP, Sakariah KK (2001) Antioxidant activity of grapeseed (Vitis vinifera) extracts on peroxidation models in vitro. Food Chem 73(3):285–290
375. Misico RI, Song LL, Veleiro AS et al (2002) Induction of quinone reductase by withanolides. J Nat Prod 65(5):677–680. doi:10.1021/np0106337

376. Bellila A, Tremblay C, Pichette A et al (2011) Cytotoxic activity of withanolies isolated from Tunisian Datura metel L. Phytochemistry 72(16):2031–2036. doi:10.1016/j.phytochem.2011.07.009

377. Piacente S, Montoro P, Oleszek W et al (2004) Yucca schidigera bark: phenolic constituents and antioxidant activity. J Nat Prod 67(5):882–885. doi:10.1021/np030369c

378. Cuéllar MJ, Giner RM, Carmen Recio M et al (1997) Zanhasaponins A and B, antiphospholipase A$_2$ saponins from an antiinflammatory extract of Zanha africana root bark. J Nat Prod 60(11):1158–1160. doi:10.1021/np970221r

379. Runyoro DKB, Kamuhabwa A, Ngassapa OD et al (2005) Cytotoxic activity of some Tanzanian medicinal plants. East Central Afr J Pharm Sci 8(2):35–39

380. Masuda T (1997) Chapter 18: Anti-inflammatory antioxidants from tropical Zingiberaceae plants isolation and synthesis of new curcuminoids. In: Sara JR, Chi-Tang H (eds) Spices: flavor chemistry and antioxidant properties, vol 18, ACS symposium series. American Chemical Society, Washington, DC, pp 219–233. doi:10.1021/bk-1997-0660.ch018

381. Fujiwara Y, Hayashida A, Tsurushima K et al (2011) Triterpenoids isolated from Zizyphus jujuba inhibit foam cell formation in macrophages. J Agric Food Chem 59(9):4544–4552. doi:10.1021/jf200193r

382. Huang X, Kojima-Yuasa A, Norikura T, Kennedy DO, Hasuma T, Matsui-Yuasa I (2007) Mechanism of the anti-cancer activity of Zizyphus jujuba in HepG2 cells. Am J Chin Med 35(3):517–532

383. Mahajan RT, Chopda MZ (2009) Phyto-pharmacology of Ziziphus jujuba Mill – a plant review. Pharmacogn Review 3(6):320–329

384. Akinyele BO, Odiyi AC (2007) Comparative study of the vegetative morphology and the existing taxonomic status of Aloe vera L. J Plant Sci 2(5):558–563. doi:10.3923/jps.2007.558.563

385. Ernst E (2000) Adverse effects of herbal drugs in dermatology. Br J Dermatol 143(5):923–929. doi:10.1046/j.1365-2133.2000.03822.x

386. Boudreau MD, Beland FA (2006) An evaluation of the biological and toxicological properties of Aloe Barbadensis (Miller), Aloe Vera. J Environ Sci Health C 24:103–154

387. Vogler BK, Ernst E (1999) Aloe vera: a systematic review of its clinical effectiveness. Br J Gen Pract 49(447):823–828

388. Gong M, Wang F, Chen Y (2002) Study on application of arbuscular-mycorrhizas in growing seedings of Aloe vera. J Chin Med Mater 25(1):1–3 (in Chinese)

389. Rossello JA, Cosín R, Boscaiu M et al (2006) Intragenomic diversity and phylogenetic systematics of wild rosemaries (Rosmarinus officinalis L. s.l., Lamiaceae) assessed by nuclear ribosomal DNA sequences (ITS). Plant Syst Evol 262(1–2):1–12. doi:10.1007/s00606-006-0454-5

390. Calabrese V, Scapagnini G, Catalano C et al (2000) Biochemical studies of a natural antioxidant isolated from rosemary and its application in cosmetic dermatology. Int J Tissue React 22(1):5–13

391. Tall JM, Seeram NP, Zhao C, Nair MG, Meyer RA, Raja SN (2004) Tart cherry anthocyanins suppress inflammation-induced pain behavior in rat. Behav Brain Res 153(1):181–188. doi:10.1016/j.bbr.2003.11.011

392. Haddad JJ, Ghadieh RM, Hasan HA, Nakhal YK, Hanbali LB (2013) Measurement of antioxidant activity and antioxidant compounds under versatile extraction conditions: II. The immuno-biochemical antioxidant properties of Black Sour Cherry (Prunus cerasus) extracts. Antiinflamm Antiallergy Agents Med Chem 12(3):229–245

393. Wang SY, Yang CW, Liao JW, Zhen WW, Chu FH, Chang ST (2008) Essential oil from leaves of Cinnamomum osmophloeum acts as a xanthine oxidase inhibitor and reduces the serum uric acid levels in oxonate-induced mice. Phytomedicine 15(11):940. doi:10.1016/j.phymed.2008.06.002

394. Anderberg A (1991) Taxonomy and phylogeny of tribe Inuleae (Asteraceae). Plant Syst Evol 176(1–2):75–123. doi:10.1007/BF00937947

395. Abid R, Qaiser M (2003) Chemotoxonomic study of Inula L. (s.str.) and its allied genera (Inuleae – Compositae) from Pakistan and Kashmir. Pak J Bot 35(2):127–140

396. Wei F, Ma LY, Jin WT et al (2004) Anti-inflammatory triterpenoid saponins from the seeds of Aesculus chinensis. Chem Pharm Bull 52(10):1246–1248. doi:10.1248/cpb.52.1246

397. Huang J, Long C (2007) Coptis teeta-based agroforestry system and its conservation potential: a case study from northwest Yunnan. AMBIO 36(4):343–349. doi:10.1579/0044-7447(2007) 36

398. Rosito MA (1975) Enumeration of the plants of Honduras. Ceiba 19(1):1–118

399. Zerega NJC, Ragone D, Motley TJ (2004) The complex origins of breadfruit (Artocarpus altilis, Moraceae): implications for human migrations in Oceania. Am J Bot 91(5):760–766. doi:10.3732/ajb.91.5.760

400. Mastelić J, Politeo O, Jerković I (2008) Contribution to the analysis of the essential oil of Helichrysum italicum (Roth) G. Don. determination of ester bonded acids and phenols. Molecules 13(4):795–803. doi:10.3390/molecules13040795

401. Patel MS, Antala BV, Barua CC, Lahkar M (2013) Anxiolytic activity of aqueous extract of Garcinia indica in mice. Int J Green Pharm 7(4):332–335. doi:10.4103/0973-8258.122089

402. Manns U, Bremer B (2010) Towards a better understanding of intertribal relationships and stable tribal delimitations within Cinchonoideae s.s. (Rubiaceae). Mol Phylogenet Evol 56(1):21–39. doi:10.1016/j.ympev.2010.04.002

403. Mohd ZZ, Abdul HA, Osman A, Saari N, Misran A (2007) Isolation and identification of antioxidative compound from fruit of Mengkudu (Morinda citrifolia L.). Int J Food Prop 10(2):363–373. doi:10.1080/10942910601052723

404. Wang MY, West BJ, Jensen CJ, Nowicki D, Su C, Palu AK, Anderson G (2002) Morinda citrifolia (Noni): a literature review and recent advances in Noni research. Pharm Sin 23(12):1127–1141

405. Won II, Renner SS (2005) The internal transcribed spacer of nuclear ribosomal DNA in the gymnosperm Gnetum. Mol Phylogenet Evol 36:581–597. doi:10.1016/j.ympev.2005.03.011

406. Won H, Renner SS (2006) Dating dispersal and radiation in the gymnosperm Gnetum (Gnetales) – clock calibration when outgroup relationships are uncertain. Syst Biol 55(4):610–622. doi:10.1080/10635150600812619

407. Hyam R, Pankhurst RJ (1995) Plants and their names: a concise dictionary. Oxford University Press, Oxford, p 515

408. Vogl S, Picker P, Mihaly-Bison J, Fakhrudin N, Atanasov AG, Heiss EH, Wawrosch C, Reznicek G, Dirsch VM, Saukel J, Kopp B (2013) Ethnopharmacological in vitro studies on Austria's folk medicine – an unexplored lore in vitro anti-inflammatory activities of 71 Austrian traditional herbal drugs. J Ethnopharmacol. doi:S0378-8741(13)00410-8. 10.1016/j.jep.2013.06.007

409. Sheeja K, Shihab PK, Kuttan G (2006) Antioxidant and anti-inflammatory activities of the plant Andrographis paniculata Nees. Immunopharmacol Immunotoxicol 28(1):129–140. doi:10.1080/08923970600626007

410. Hoot SB, Meyer KM, Manning JC (2012) Phylogeny and reclassification of Anemone (Ranunculaceae), with an emphasis on austral species. Syst Bot 37(1):139–152

411. Singh SS, Pandey SC, Srivastava S et al (2003) Chemistry and medicinal properties of Tinospora cordifolia. Indian J Pharmacol 35:83–91

412. Ozaki Y, Kawahara N, Harada M (1991) Anti-inflammatory effect of Zingiber cassumunar Roxb. And its active principles. Chem Pharm Bull 39(9):2353–2356

413. White OE, Bowden WM (1947) Oriental and American bittersweet hybrids. J Hered 38(4):125–128

414. Iannetta PPM, Wyman M, Neelam A, Jones C, Taylor MA, Davies HV, Sexton R (2000) A causal role for ethylene and endo-beta-1, 4-glucanase in the abscission of red-raspberry (Rubus idaeus) drupelets. Physiol Plant 110(4):535–543. doi:10.1111/j.1399-3054.2000.1100417.x

415. Liu M, Li XQ, Weber C, Lee CY, Brown J, Liu RH (2002) Antioxidant and antiproliferative activities of raspberries. J Agric Food Chem 50(10):2926–2930. doi:10.1021/jf0111209

416. Heinonen M (2007) Antioxidant activity and antimicrobial effect of berry phenolics a Finnish perspective. Mol Nutr Food Res 51(6):684–691. doi:10.1002/mnfr.200700006

417. Cerdá B, Tomás-Barberán FA, Espín JC (2005) Metabolism of antioxidant and chemopreventive ellagitannins from strawberries, raspberries, walnuts, and oak-aged wine in humans: identification of biomarkers and individual variability. J Agric Food Chem 53(2):227–235. doi:10.1021/jf049144d
418. Vokou D, Kokkini S, Bessière JM (1988) Origanum onites (Lamiaceae) in Greece: distribution, volatile oil yield, and composition. Econ Bot 42(3):407–412. doi:10.1007/BF02860163
419. Sarac N, Ugur A (2008) Antimicrobial activities of the essential oils of Origanum onites L., Origanum vulgare L. Subspecies hirtum (Link) Ietswaart, Satureja thymbra L., and Thymus cilicicus Boiss. & Bal. Growing wild in Turkey. J Med Food 11(3):568–573. doi:10.1089/jmf.2007.0520
420. Arulselvan P, Senthilkumar GP, Sathish Kumar D, Subramanian S (2006) Anti-diabetic effect of Murraya koenigii leaves on streptozotocin induced diabetic rats. Pharmazie 61(10):874–877
421. Henrotin Y, Clutterbuck AL, Allaway D et al (2010) Biological actions of curcumin on articular chondrocytes. Osteoarthr Cartil 18(2):141–149. doi:10.1016/j.joca.2009.10.002
422. Nagpal M, Sood S (2013) Role of curcumin in systemic and oral health: an overview. J Nat Sci Biol Med 4(1):3–7. doi:10.4103/09769668.107253
423. Chattopadhyay I, Biswas K, Bandyopadhyay U, Banerjee RK (2004) Turmeric and curcumin: biological actions and medicinal applications. Curr Sci 87(1):44–53
424. Pardo F, Perich F, Villarroel L, Torres R (1993) Isolation of verbascoside, an antimicrobial constituent of Buddleja globosa leaves. J Ethnopharmacol 39(3):221–222. doi:10.1016/0378-8741(93)90041-3
425. Backhouse N, Rosales L, Apablaza C et al (2008) Analgesic, anti-inflammatory and antioxidant properties of Buddleja globosa, Buddlejaceae. J Ethnopharmacol 116(2):263–269. doi:10.1016/j.jep.2007.11.025
426. Houghton P (1996) Buddlejone, a diterpene from Buddleja albiflora. Phytochemistry 42(2):485–488. doi:10.1016/0031-9422(96)00001-5
427. Pareek A, Suthar M, Rathore GS, Bansal V (2011) Feverfew (Tanacetum parthenium L.): a systematic review. Pharmacogn Rev 5(9):103–110. doi:10.4103/0973-7847.79105
428. Guzman ML, Rossi RM, Karnischky L et al (2005) The sesquiterpene lactone parthenolide induces apoptosis of human acute myelogenous leukemia stem and progenitor cells. Blood 105(11):4163–4169. doi:10.1182/blood-2004-10-4135
429. Draves AH, Walker SE (2004) Parthenolide content of Canadian commercial feverfew preparations: label claims are misleading in most cases. Can Pharm J 136(10):23–30
430. Seeram NP (2008) Berry fruits: compositional elements, biochemical activities, and the impact of their intake on human health, performance, and disease. J Agric Food Chem 56(3):627–629. doi:10.1021/jf071988k
431. Kapasakalidis PG, Rastall RA, Gordon MH (2006) Extraction of polyphenols from processed black currant (Ribes nigrum L.) residues. J Agric Food Chem 54(11):4016–4021. doi:10.1021/jf0529991
432. Mcdougall GJ, Gordon S, Brennan R, Stewart D (2005) Anthocyanin-flavanol condensation products from black currant (Ribes nigrum L.). J Agric Food Chem 53(20):7878–7885. doi:10.1021/jf0512095
433. Vogl S, Picker P, Mihaly-Bison J, Fakhrudin N et al (2013) Ethnopharmacological in vitro studies on Austria's folk medicine-An unexplored lore in vitro anti-inflammatory activities of 71 Austrian traditional herbal drugs. J Ethnopharmacol 149(3):750–771. doi:10.1016/j.jep.2013.06.007
434. Traitler H, Winter H, Richli U, Ingenbleek Y (1984) Characterization of gamma-linolenic acid in Ribes seed. Lipids 19(12):923–928. doi:10.1007/BF02534727
435. Nicolosi E, Deng ZN, Gentile A, La Malfa S, Continella G, Tribulato E (2000) Citrus phylogeny and genetic origin of important species as investigated by molecular markers. TAG Theor Appl Genet 100(8):1155–1166. doi:10.1007/s001220051419

436. Duan JA, Wang LY, Qian SH, Su SL, Tang YP (2008) A new cytotoxic prenylated dihydro-benzofuran derivative and other chemical constituents from the rhizomes of Atractylodes lancea DC. Arch Pharm Res 12(8):965–969. doi:10.1007/s12272-001-1252- z

437. John MM, Jeffery LD (2000) Signal transduction in the plant immune response. Trends Biochem Sci 12(2):79–82. doi:10.1016/S0968-0004(99)01532-7

438. Nojiri H, Sugimori M, Yamane H, Nishimura Y, Yamada A, Shibuya N, Kodama O, Murofushi N, Omori T (1996) Involvement of jasmonic acid in elicitor-induced phytoalexin production in suspension-cultured rice cells. Plant Physiol 12(2):387–392

439. Dincer C, Karaoglan M, Erden F, Tetik N, Topuz A, Ozdemir F (2011) Effects of baking and boiling on the nutritional and antioxidant properties of sweet potato [Ipomoea batatas (L.) Lam.] cultivars. Plant Foods Hum Nutr 66(4):341–347. doi:10.1007/s11130-011-0262-0

440. Chen ZL (1987) The acetylenes from Atractylodes macrocephala. Planta Med 53:493–494

441. Weng CJ, Fang PS, Chen DH, Chen KD, Yen GC (2010) Anti-invasive effect of a rare mush-room, Ganoderma colossum, on human hepatoma cells. J Agric Food Chem 58(13):7657–7663. doi:10.1021/jf101464h

442. Kirk PM, Cannon PF, Minter DW, Stalpers JA (2008) Dictionary of the fungi, 10th edn. CABI, Wallingford, p 272

443. Yuen JW, Gohel MD (2005) Anticancer effects of Ganoderma lucidum: a review of scientific evidence. Nutr Cancer 53(1):11–17. doi:10.1207/s15327914nc5301_2

444. Takashima J, Ohsaki A (2002) Brosimacutins A-I, nine new flavonoids from Brosimum acu-tifolium. J Nat Prod 65(12):1843–1847

445. Bouskela E, Cyrino FZGA, Marcelon G (1993) Effects of Ruscus extract on the internal diameter of arterioles and venules of the hamster cheek pouch microcirculation. J Cardiovasc Pharmacol 22(2):221–224. doi:10.1097/00005344-199308000-00008

446. MacKay D (2001) Hemorrhoids and varicose veins: a review of treatment options. Altern Med Rev 6(2):126–140

447. Harbowy ME, Balentine DA, Davies AP, Cai Y (1997) Tea chemistry. Crit Rev Plant Sci 16(5):415–480

448. Toomey VM, Nickum EA, Flurer CL (2012) Cyanide and amygdalin as indicators of the pres-ence of bitter almonds in imported raw almonds. J Forensic Sci 57(5):1313–1317. doi:10.1111/j.1556-4029.2012.02138.x

449. Adhvaryu MR, Reddy MN, Vakharia BC (2008) Prevention of hepatotoxicity due to anti tuberculosis treatment: a novel integrative approach. World J Gastroenterol 14(30):4753–4762

450. Franco LA, Matiz GE, Calle J, Pinzón R, Ospina LF (2007) Anti-inflammatory activity of extracts and fractions obtained from Physalis peruviana L. calyces. Biomedica 27(1):110–115

451. Caamal-Maldonado JA, Jimenez JJ, Torres A, Anaya A (2001) The use of allelopathic legume cover and mulch species for weed control in cropping systems. Agron J 93(1):27–36

452. Shemesh A, Mayo WL (1991) Australian tea tree oil: a natural antiseptic and fungicidal agent. Aust J Pharm 72:802–803

453. Hammer K, Carson C, Riley T, Nielsen J (2006) A review of the toxicity of Melaleuca alter-nifolia (tea tree) oil. Food Chem Toxicol 44(5):616–625. doi:10.1016/j.fct.2005.09.001

454. Blanco MM, Costa CA, Freire AO, Santos JG, Costa M (2009) Neurobehavioral effect of essential oil of Cymbopogon citratus in mice. Phytomedicine 16(2–3):265–270. doi:10.1016/j.phymed.2007.04.007

455. Samuelsen AB (2000) The traditional uses, chemical constituents and biological activities of Plantago major L. A review. J Ethnopharmacol 77(1–2):1. doi:10.1016/S0378-8741(00)00212-9

456. Ngamkitidechakul C, Jaijoy K, Hansakul P, Soonthornchareonnon N, Sireeratawong S (2010) Antitumour effects of phyllanthus emblica L.: induction of cancer cell apoptosis and Inhibition of in vivo tumour promotion and in vitro invasion of human cancer cells. Phytother Res 24(9):1405–1413. doi:10.1002/ptr.3127

457. Sidhu S, Pandhi P, Malhotra S, Vaiphei K, Khanduja KL (2011) Beneficial effects of Emblica officinalisinl-arginine-induced acute pancreatitis in rats. J Med Food 14(1–2):147–155. doi:10.1089/jmf.2010.1108
458. Rao TP, Sakaguchi N, Juneja LR, Wada E, Yokozawa T (2005) Amla (Emblica officinalis Gaertn.) extracts reduce oxidative stress in streptozotocin-induced diabetic rats. J Med Food 8(3):362–368. doi:10.1089/jmf.2005.8.362
459. Yoshikawa M, Uchida E, Kawaguchi A, Kitagawa I, Yamahara J (1992) Galloyl-oxypaeoniflorin, suffruticosides A, B, C, and D, five new antioxidative glycosides, and suffruticoside E, A paeonol glycoside, from Chinese moutan cortex. Chem Pharm Bull 40(8):2248–2250
460. Clauson KA, Shields KM, McQueen CE, Persad N (2003) Safety issues associated with commercially available energy drinks. J Am Pharm Assoc 48(3):e55–e63. doi:10.1331/JAPhA.2008.07055
461. Qi LW, Wang CZ, Yuan CS (2011) Ginsenosides from American ginseng: chemical and pharmacological diversity. Phytochemistry 72(8):689–699. doi:10.1016/j.phytochem.2011.02.012
462. Cichoke AJ (2001) Secrets of Native American herbal remedies: a comprehensive guide to the Native American tradition of using herbs and the mind/body/spirit connection for improving health and well-being. Avery/Penguin Putnam, New York
463. Rao KV, Kasanah N, Wahyuono S et al (2004) Three new manzamine alkaloids from a common Indonesian sponge and their activity against infectious and tropical parasitic diseases. J Nat Prod 67(8):1314–1318. doi:10.1021/np0400095
464. Zhang B, Higuchi R, Miyamoto T et al (2008) Neuritogenic activity-guided isolation of a free base form manzamine A from a marine sponge, Acanthostrongylophora aff. ingens (Thiele, 1899). Chem Pharm Bull 56(6):866–869
465. Meragelman KM, West LM, Northcote PT et al (2002) Unusual sulfamate indoles and a novel indolo[3,2-a]carbazole from Ancorina sp. J Org Chem 67(19):6671–6677. doi:10.1021/jo020120k
466. Simon-Levert A, Arrault A, Bontemps-Subielos N, Canal C, Banaigs B (2005) Meroterpenes from the Ascidian Aplidium aff. Densum. J Nat Prod 68(9):1412–1415. doi:10.1021/np050110p
467. Li G-Y, Li B-G, Yang T et al (2005) Sesterterpenoids, terretonins A-D, and an alkaloid aster-relenin from Aspergillus terreus. J Nat Prod 68(8):1243–1246. doi:10.1021/np0501738
468. Subazini TK, Ramesh Kumar G (2011) Characterization of Lovastatin biosynthetic cluster proteins in Aspergillus terreus strain ATCC 20542. Bioinformation 6(7):250–254
469. Elsebai MF, Rempel V, Schnakenburg G, Kehraus S, Müller CE, König GM (2011) Identification of a potent and selective cannabinoid CB_1 receptor antagonist from Auxarthron reticulatum. ACS Med Chem Lett 2(11):866–869. doi:10.1021/ml200183z
470. Sharma V, Lansdell TA, Jin G et al (2004) Inhibition of cytokine production by hymenialdisine derivatives. J Med Chem 47(14):3700–3703. doi:10.1021/jm040013d
471. Sharma V, Lansdell TA, Jin G et al (2004) Inhibition of cytokine production by hymenialdisine derivatives. J Med Chem 47(14):3700–3703
472. Wu S-L, Sung P-J, Su J-H et al (2003) Briaexcavatolides S-V, four new briaranes from a Formosan gorgonian Briareum excavatum. J Nat Prod 66(9):1252–1256. doi:10.1021/np030102d
473. Yeh T-T, Wang S-K, Dai C-F et al (2012) Briacavatolides A-C, new briaranes from the Taiwanese octocoral Briareum excavatum. Mar Drugs 10(5):1019–1026. doi:10.3390/md10051019
474. Sheu J-H, Sung P-J, Su J-H et al (1999) Excavatolides U-Z, new briarane diterpenes from the Gorgonian Briareum excavatum. J Nat Prod 62(10):1415–1420. doi:10.1021/np990302i
475. Sung P-J, Su J-H, Wang G-H et al (1999) Excavatolides F-M, new briarane diterpenes from the Gorgonian Briareum excavatum. J Nat Prod 62(3):457–463. doi:10.1021/np980446h

476. Sheu J-H, Sung P-J, Cheng M-C et al (1998) Novel cytotoxic diterpenes, excavatolides A-E, isolated from the Formosan gorgonian Briareum excavatum. J Nat Prod 61(5):602–608. doi:10.1021/np970553w

477. Kwak JH, Schmitz FJ, Williams GC (2001) Milolides, new briarane diterpenoids from the western Pacific octocoral Briareum stechei. J Nat Prod 64(6):754–760. doi:10.1021/np010009u

478. Appleton DR, Sewell MA, Berridge MV et al (2002) A new biologically active malyngamide from a New Zealand collection of the sea hare Bursatella leachii. J Nat Prod 65(4):630–631. doi:10.1021/np010511e

479. Zampella A, D'Auria MV, Paloma LG et al (1996) Callipeltin A, an Anti-HIV cyclic depsipeptide from the new Caledonian Lithistida sponge Callipelta sp. J Am Chem Soc 118(26):6202–6209. doi:10.1021/ja954287p

480. Tan LT, Williamson RT, Gerwick WH (2000) cis, cis- and trans, trans-ceratospongamide, new bioactive cyclic heptapeptides from the Indonesian red alga Ceratodictyon spongiosum and symbiotic sponge Sigmadocia symbiotica. J Org Chem 65(2):419–425. doi:10.1021/jo991165x

481. Akiyama T, Ueoka R, van Soest RW et al (2009) Ceratodictyols, 1-glyceryl ethers from the red alga-sponge association Ceratodictyon spongiosum/Haliclona cymaeformis. J Nat Prod 72(8):1552–1554. doi:10.1021/np900355m

482. Tomono Y, Hirota H, Fusetani N (1999) Isogosterones A-D, antifouling 13,17-secosteroids from an Octocoral Dendronephthya sp. J Org Chem 64(7):2272–2275. doi:10.1021/jo981828v

483. Harder T, Lau SC, Dobretsov S, Fang TK, Qian PY (2003) A distinctive epibiotic bacterial community on the soft coral Dendronephthya sp. and antibacterial activity of coral tissue extracts suggest a chemical mechanism against bacterial epibiosis. FEMS Microbiol Ecol 43(3):337–347

484. Golik J, Dickey JK, Todderud G et al (1997) Isolation and structure determination of sulfono-quinovosyl dipalmitoyl glyceride, a P-selectin receptor inhibitor from the alga Dictyochloris fragrans. J Nat Prod 60(4):387–389. doi:10.1021/np9606761

485. Pedpradab S, Edrada RA, Ebel R et al (2004) New β-carboline alkaloids from the Andaman Sea Sponge Dragmacidon sp. J Nat Prod 67(12):2113–2116. doi:10.1021/np0401516

486. Hooper GJ, Davies-Coleman MT, Schleyer M (1997) New diterpenes from the South African soft coral Eleutherobia aurea. J Nat Prod 60(9):889–893. doi:10.1021/np970180z

487. Jensen PR, Fenical W (2005) New natural-product diversity from marine actinomycetes. In: Zhang L, Demain AL (eds) Natural products: drug discovery and therapeutic medicine. Humana Press Inc, Totowa, p 315

488. Shi Y-P, Rodríguez AD, Padilla OL (2001) Calyculaglycosides D and E, novel cembrane glycosides from the Caribbean gorgonian octocoral Eunicea species and structural revision of the aglycon of calyculaglycosides A-C. J Nat Prod 64(11):1439–1443. doi:10.1021/np0104121

489. Garzón SP, Rodríguez AD, Sánchez JA et al (2005) Sesquiterpenoid metabolites with anti-plasmodial activity from a Caribbean gorgonian coral, Eunicea sp. J Nat Prod 68(9):1354–1359

490. De Rosa S, Crispino A, De Giulio A et al (1998) A new cacospongionolide inhibitor of human secretory phospholipase $_{A2}$ from the Tyrrhenian sponge Fasciospongia cavernosa and absolute configuration of cacospongionolides. J Nat Prod 61(7):931–935. doi:10.1021/np980122t

491. De Rosa S, Crispino A, De Giulio A et al (1999) A new cacospongionolide derivative from the sponge Fasciospongia cavernosa. J Nat Prod 62(9):1316–1318. doi:10.1021/np9901251

492. Venkateswarlu Y, Farooq Biabani MA (1994) A new trisnorditerpene from the sponge Fasciospongia cavernosa. J Nat Prod 57(11):1578–1579. doi:10.1021/np50113a019

493. Fontana A, Cavaliere P, Ungur N et al (1999) New scalaranes from the nudibranch Glossodoris atromarginata and its sponge Prey. J Nat Prod 62(10):1367–1370. doi:10.1021/np9900932

494. Fontana A, Mollo E, Ortea J et al (2000) Scalarane and homoscalarane compounds from the nudibranchs Glossodoris sedna and Glossodoris dalli: chemical and biological properties. J Nat Prod 63(4):527–530. doi:10.1021/np990506z

495. Fontana A, Cavaliere P, Ungur N et al (1999) New scalaranes from the nudibranch Glossodoris atromarginata and its sponge Prey. J Nat Prod 62:1367–1370

496. Machmudah S, Shotipruk A, Goto M et al (2006) Extraction of astaxanthin from Haematococcus pluvialis using supercritical CO_2 and ethanol as entrainer. Ind Eng Chem Res 45(10):3652–3657. doi:10.1021/ie051357k

497. Ryu G, Matsunaga S, Fusetani N (1996) Three new cytotoxic sesterterpenes from the marine sponge Hyrtios cf. erectus. J Nat Prod 59(5):515–517. doi:10.1021/np960130e

498. Youssef DTA, Yamaki RK, Kelly M et al (1995) A novel cytotoxic sesterterpene from the red sea sponge Hyrtios erecta. J Nat Prod 65(1):2–6. doi:10.1021/np0101853

499. Pettit RK, McAllister SC, Pettit GR, Herald CL, Johnson JM, Cichacz ZA (1997) A broad-spectrum antifungal from the marine sponge Hyrtios erecta. Int J Antimicrob Agents 9(3):147–152

500. Kirsch G, König GM, Anthony D et al (2000) A new bioactive sesterterpene and antiplasmo-dial alkaloids from the marine sponge Hyrtios cf. erecta. J Nat Prod 63(6):825–829. doi:10.1021/np990555b

501. Miyaoka H, Nishijima S, Mitome H, Yamada Y (2000) Three new scalarane sesterterpenoids from the Okinawan sponge Hyrtios erectus. J Nat Prod 63(10):1369–1372. doi:10.1021/np000115g

502. Pettit GR, Butler MS, Williams MD et al (1996) Isolation and structure of hemibastadinols 1-3 from the papua new guinea marine sponge Ianthella bast. J Nat Prod 59(10):927–934. doi:10.1021/np960249n

503. Brunner E, Ehrlich H, Schupp P et al (2009) Chitin-based scaffolds are an integral part of the skeleton of the marine demosponge Ianthella basta. J Struct Biol 168:539–547

504. Franklin MA, Penn SG, Lebrilla CB et al (1996) Bastadin 20 and bastadin O-sulfate esters from Ianthella basta: novel modulators of the Ry1R FKBP12 receptor complex. J Nat Prod 59(12):1121–1127. doi:10.1021/np960507g

505. Ortlepp S (2008) Bastadins and related compounds from the marine sponges Ianthella basta and Callyspongia sp: structure elucidation and biological activities. Cuvillier Verlag, Gottingen

506. Greve H, Meis S, Kassack MU et al (2007) New iantherans from the marine sponge Ianthella quadrangulata: novel agonists of the $P2Y_{11}$ receptor. J Med Chem 50(23):5600–5607. doi:10.1021/jm070043r

507. Greve H, Kehraus S, Krick A, Kelter G, Maier A, Fiebig HH, Wright AD, König GM (2008) Cytotoxic bastadin 24 from the Australian sponge Ianthella quadrangulata. J Nat Prod 71(3):309–312. doi:10.1021/np070373e

508. García M, Rodríguez J, Jiménez C (1999) Absolute structures of new briarane diterpenoids from Junceella fragilis. J Nat Prod 62(2):257–260. doi:10.1021/np980331d

509. Tsai S, Spikings E, Huang IC, Lin C (2011) Study on the mitochondrial activity and mem-brane potential after exposing later stage oocytes of two gorgonian corals (Junceella juncea and Junceella fragilis) to cryoprotectants. Cryo Lett 32(1):1–12

510. Shen Y-C, Lin Y-C, Ko C-L et al (2003) New briaranes from the Taiwanese gorgonian Junceella juncea. J Nat Prod 66(2):302–305. doi:10.1021/np0203584

511. Qi SH, Zhang S, Qian PY et al (2012) Antifeedant and antifouling briaranes from the South China Sea gorgonian Junceella juncea. Chem Nat Compd 45(1):49–54. doi:10.1007/s10600-009-9255-8

512. Matthée GF, König GM, Wright AD (1998) Three new diterpenes from the marine soft coral Lobophytum crassum. J Nat Prod 61(2):237–240. doi:10.1021/np970458n

513. Lin S-T, Wang S-K, Duh C-Y (2011) Cembranoids from the Dongsha Atoll soft coral Lobophytum crassum. Mar Drugs 9(12):2705–2716. doi:10.3390/md9122705

514. Jaki B, Orjala J, Sticher O (1999) A novel extracellular diterpenoid with antibacterial activity from the cyanobacterium Nostoc commune. J Nat Prod 62(3):502–503

515. Iwasaki J, Ito H, Aoyagi M et al (2006) Briarane-type diterpenoids from the Okinawan soft coral Pachyclavularia violacea. J Nat Prod 69(1):2–6. doi:10.1021/np0580661

516. Ponomarenko LP, Kalinovsky AI, Stonik VA (2004) New scalarane-based sesterterpenes from the sponge Phyllospongia madagascarensis. J Nat Prod 67(9):1507–1510. doi:10.1021/np040073m

517. Cuéllar MJ, Giner RM, Recio MC et al (1996) Two fungal lanostane derivatives as phospholipase A₂ inhibitors. J Nat Prod 59(10):977–979. doi:10.1021/np9604339

518. Li GH, Shen YM, Zhang KQ (2005) Nematicidal activity and chemical component of Poria cocos. J Microbiol 43(1):17–20

519. Rodríguez AD, Shi J-G, Huang SD (1999) Highly oxygenated pseudopterane and cembranolide diterpenes from the Caribbean sea feather Pseudopterogorgia bipinnata. J Nat Prod 62(9):1228–1237. doi:10.1021/np990064r

520. Ospina CA, Rodríguez AD, Sánchez JA et al (2005) Caucanolides A–F, unusual antiplasmodial constituents from a colombian collection of the gorgonian coral Pseudopterogorgia bipinnata. J Nat Prod 68(10):1519–1526

521. Rodríguez AD, Ramírez C, Rodríguez II (1999) Elisabatins A and B: new amphilectane-type diterpenes from the West Indian sea whip Pseudopterogorgia elisabethae. J Nat Prod 62(7):997–999. doi:10.1021/np990090p

522. Look SA, Fenical W, Jacobs RS, Clardy J (1986) The pseudopterosins: anti-inflammatory and analgesic natural products from the sea whip Pseudopterogorgia elisabethae. Proc Natl Acad Sci 83(17):6238–6240

523. Rodríguez AD, González E, Huang SD (1998) Unusual terpenes with novel carbon skeletons from the West Indian sea whip Pseudopterogorgia elisabethae (Octocorallia). J Org Chem 63(20):7083–7091. doi:10.1021/jo981385v

524. Rodríguez AD, Ramírez C, Rodríguez II et al (2000) Novel terpenoids from the West Indian sea whip Pseudopterogorgia elisabethae (Bayer). Elisapterosins A and B: rearranged diterpenes possessing an unprecedented cagelike framework. J Org Chem 65(5):1390–1398. doi:10.1021/jo9914869

525. Marrero J, Benítez J, Rodríguez AD et al (2008) Bipinnatins K–Q, minor cembrane-type diterpenes from the West Indian Gorgonian Pseudopterogorgia kallos: isolation, structure assignment and evaluation of biological activities. J Nat Prod 71(3):381–389. doi:10.1021/np0705561

526. Barsby T, Kubanek J (2005) Isolation and structure elucidation of feeding deterrent diterpenoids from the Sea Pansy, Renilla reniformis. J Nat Prod 68(4):511–516. doi:10.1021/np049609u

527. Srikantha T, Klapach A, Lorenz WW et al (1996) The sea pansy Renilla reniformis luciferase serves as a sensitive bioluminescent reporter for differential gene expression in Candida albicans. J Bacteriol 178(1):121–129

528. Casapullo A, Giuseppe B, Ines B et al (2000) New bisindole alkaloids of the topsentin and hamacanthin classes from the Mediterranean marine sponge Rhaphisia lacazei. J Nat Prod 63(4):447–451. doi:10.1021/np9903292

529. Zhang C, Li J, Su J et al (2006) Cytotoxic diterpenoids from the soft coral Sarcophyton crassocaule. J Nat Prod 69(10):1476–1480. doi:10.1021/np050499g

530. Lin W-Y, Lu Y, Su J-H et al (2011) Bioactive cembranoids from the dongsha atoll soft coral Sarcophyton crassocaule. Mar Drugs 9(6):994–1006. doi:10.3390/md9060994

531. König GM, Wright AD (1998) New cembranoid diterpenes from the soft coral Sarcophyton ehrenbergi. J Nat Prod 61(4):494–496. doi:10.1021/np9704112

532. Wang S-K, Hsieh M-K, Duh C-Y (2012) Three new cembranoids from the Taiwanese Soft Coral Sarcophyton ehrenbergi. Mar Drugs 10(7):1433–1444. doi:10.3390/md10071433

533. Kuo YH, Hsu HC, Chen YC (2012) A novel compound with antioxidant activity produced by Serratia ureilytica TKU013. J Agric Food Chem 60(36):9043–9047. doi:10.1021/jf302481n

534. Renner MK, Shen Y-C, Cheng X-C et al (2005) Cyclomarins A–C, new antiinflammatory cyclic peptides produced by a marine bacterium (Streptomyces sp.). J Am Chem Soc 121(49):11273–11276. doi:10.1021/ja992482o

535. Pereira R, Medeiros YS, Fröde TS (2006) Antiinflammatory effects of Tacrolimus in a mouse model of pleurisy. Transpl Immunol 16(2):105–111

536. Sudha S, Selvam M (2011) Antibacterial activity of a new Streptomyces sp. SU isolated from Rhizosphere soil. J Pharm Res 4(5):1515–1516

537. Mohammed R, Peng J, Kelly M et al (2006) Cyclic heptapeptides from the Jamaican sponge Stylissa caribica. J Nat Prod 69(12):1739–1744. doi:10.1021/np060006n

538. Buchanan MS, Carroll AR, Addepalli R et al (2007) Natural products, stylissadines A and B, specific antagonists of the P2X$_7$ receptor, an important inflammatory target. J Org Chem 72(7):2309–2317. doi:10.1021/jo062007q

539. Prinsep MR, Thomson RA (1996) Tolypodiol: an antiinflammatory diterpenoid from the cyanobacterium Tolypothrix nodosa. J Nat Prod 59(8):786–788. doi:10.1021/np9602574

540. Prinsep MR, Caplan FR, Moore RE et al (1992) Tolyporphin, a novel multidrug resistance reversing agent from the blue-green alga Tolypothrix nodosa. J Am Chem Soc 114(1):385–387. doi:10.1021/ja00027a072

541. Horgen FD, Sakamoto B, Scheuer PJ (2000) New triterpenoid sulfates from the red alga Tricleocarpa fragilis. J Nat Prod 63(2):210–216. doi:10.1021/np990448h

542. Veluri R, Oka I, Wagner-Döbler I (2003) New indole alkaloids from the North Sea bacterium Vibrio parahaemolyticus. J Nat Prod 66(11):1520–1523. doi:10.1021/np030288g

543. Twedt RM, Novelli RE, Spaulding PL et al (1970) Comparative hemolytic activity of Vibrio parahaemolyticus and related vibrios. Infect Immun 1(4):394–399

544. Deyrup Stephen T, Gloer James B, Kerry O'D et al (2007) Kolokosides A-D: triterpenoid glycosides from a Hawaiian isolate of Xylaria sp. J Nat Prod 70(3):378–382. doi:10.1021/np060546k

545. Liu X, Dong M, Chen X, Jiang M, Lv X, Zhou J (2008) Antimicrobial activity of an endophytic Xylaria sp.YX-28 and identification of its antimicrobial compound 7-amino-4-methylcoumarin. Appl Microbiol Biotechnol 78(2):241–247

546. Hua K-F, Hsu H-Y, Su Y-C et al (2006) Study on the antiinflammatory activity of methanol extract from seagrass Zostera japonica. J Agric Food Chem 54(2):306–311. doi:10.1021/jf0509658

547. Abe M, Yokota K, Kurashima A et al (2009) High water temperature tolerance in photosynthetic activity of Zostera japonica Ascherson & Graebner seedlings from Ago Bay, Mie Prefecture, central Japan. Fish Sci 75(5):1117–1123. doi:10.1007/s12562-009-0141

548. Szakacs G, Morovjan G, Tengerdy R (1998) Production of lovastatin by a wild strain of Aspergillus terreus. Biotechnol Lett 20(4):411–415

549. Sunga P-J, Sua Y-D, Li G-Y (2009) Excavatoids A–D, new polyoxygenated briaranes from the octocoral Briareum excavatum. Tetrahedron xxx:1–7

550. Lim SC, de Voogd N, Tan KS (2008) A guide to sponges of Singapore. Singapore Science Centre, Singapore, 173

551. Bouchet P, Caballer M (2012) Doriprismatica atromarginata. World Register of Marine Species. http://www.marinespecies.org/aphia.php?p=taxdetails&id=558499

552. Rudman WB (1990) The Chromodorididae (Opisthobranchia: Mollusca) of the Indo-West Pacific: further species of Glossodoris, Thorunna and the Chromodoris aureomarginata colour group. Zool J Linnean Soc 100:263–326

553. Johnson RF, Gosliner TM (2012) Traditional taxonomic groupings mask evolutionary history: a molecular phylogeny and new classification of the chromodorid nudibranchs. PLoS ONE 7(4):e33479

554. Lorentz RT, Cysewski GR (2000) Commercial potential for Haematococcus microalgae as a natural source of astaxanthin. Trends Biotechnol 18:160–167

555. Ashour MA, Elkhayat ES, Ebel R et al (2007) Indole alkaloid from the red sea sponge Hyrtios erectus. ARKIVOC xv:225–231

556. Heckrodt TJ, Mulzer J (2005) Marine natural products from Pseudopterogorgia elisabethae: structures, biosynthesis, pharmacology, and total synthesis, natural products synthesis II. Top Curr Chem 244:1–41

557. Espada A, Rivera Sagredo A, De la Fuente JM et al (1997) New cytochalasins from the fungus Xylaria hypoxylon. Tetrahedron 53(18):6485–6492
558. Robinson SC, Laks PE (2010) Culture age and wood species affect zone line production of Xylaria polymorpha. Open Mycol J 4:18–21
559. Liu Q, Wang H, Ng TB (2006) First report of a xylose-specific lectin with potent hemagglutinating, antiproliferative and anti-mitogenic activities from a wild ascomycete mushroom. Biochim Biophys Acta 1760(12):1914–1919. doi:10.1016/j.bbagen.2006.07.010
560. Kämpfer P (2006) The family streptomycetaceae, Part I: Taxonomy. In: Dworkin M et al (eds) The prokaryotes: a handbook on the biology of bacteria. Springer, Berlin, pp 538–604
561. Labeda DP (2010) Multilocus sequence analysis of phytopathogenic species of the genus Streptomyces. Int J Syst Evol Microbiol 61(10):2525. doi:10.1099/ijs.0.028514-0
562. Dumbauld BR, Wyllie-Echeverria S (2003) The influence of burrowing thalassinid shrimps on the distribution of intertidal seagrasses in Willapa Bay, Washington, USA. Aquat Bot 77:27–42
563. Harrison PG (1982) Comparative growth of Zostera japonica Aschers. & Graebn. and Z. marina under simulated intertidal and subtidal conditions. Aquat Bot 14:373–379
564. Welch JJ (2010) The Island rule and Deep-Sea gastropods: re-examining the evidence. PLoS ONE 5(1):e8776. doi:10.1371/journal.pone.0008776
565. Tamaru Y, Takani Y, Yoshida T, Sakamoto T (2005) Crucial role of extracellular polysaccharides in desiccation and freezing tolerance in the terrestrial cyanobacterium Nostoc commune. Appl Environ Microbiol 71(11):7327–7333. doi:10.1128/AEM.71.11.7327-7333.2005
566. Wu Y, Wang D (2009) A new class of natural glycopeptides with sugar moiety-dependent antioxidant activities derived from Ganoderma lucidum fruiting nodies. J Proteome Res 8(2):436–442. doi:10.1021/pr800554w
567. Sonne C, Dietz R, Hans JS et al (2006) Impairment of cellular immunity in west Greenland sledge dogs (Canis familiaris) dietary exposed to polluted minke whale (Balaenoptera acutorostrata) blubber. Environ Sci Technol 40(6):2056–2062. doi:10.1021/es052151d
568. Brix O, Condò SG, Bardgard A et al (1990) Temperature modulation of oxygen transport in a diving mammal (Balaenoptera acutorostrata). Biochem J 271(2):509–513
569. Thwin MM, Gopalakrishnakone P, Kini RM et al (2000) Recombinant antitoxic and antiinflammatory factor from the nonvenomous snake Python reticulates: phospholipase A_2 inhibition and venom neutralizing potential. Biochemistry 39(31):9604–9611. doi:10.1021/bi000395z
570. Amira Mnari B, Harzallah HJ, Dhibi M et al (2010) Nutritional fatty acid quality of raw and cooked farmed and wild sea bream (Sparus aurata). J Agric Food Chem 58(1):507–512. doi:10.1021/jf902096w
571. Cuesta A, Esteban MA, Meseguer J (2002) Natural cytotoxic activity in seabream (Sparus aurata L.) and its modulation by vitamin C. Fish Shellfish Immunol 13(2):97–109
572. Shine R, Harlow PS, Keogh JS (1998) The influence of sex and body size on food habits of a giant tropical snake, Python reticulatus. Funct Ecol 12(2):248–258
573. Akkol K, Orhan DD, Gürbüz I, Yesilada E (2010) In vivo activity assessment of a "honey-bee pollen mix" formulation. Pharm Biol 48(3):253–259. doi:10.3109/13880200903085482
574. Engel MS (1999) The taxonomy of recent and fossil honey bees (Hymenoptera: Apidae: Apis). J Hymenopt Res 8:165–196
575. Arnason U, Gullberg A, Widegren B (1993) Cetacean mitochondrial DNA control region: sequences of all extant baleen whales and two sperm whale species. Mol Biol Evol 10(5):960–970

Abstract

Studies performed using in vivo and in vitro experimental model systems are critical components of the effort to identify the effect of new drug molecules. Experimental models permit the precise quantification of exposure levels, to eliminate many external variables that may alter drug response. Experimental studies can be designed in a manner to include specific endpoint evaluations that can generate important data concerning possible biological mechanisms of drug action. Thus, it becomes mandatory to access the cause of disease, and the mechanisms behind it, through experimental models which generally involve animals, in vitro studies, primates, and even humans to a certain extent. Interspecies differences and high-dose to low-dose extrapolations remain important challenges to the interpretation and application of experimental data to assessments of human risk. Understanding the basic mechanisms and pathogenesis of inflammation are essential for the development of new treatment approaches and therapeutic agents.

5.1 In Vitro Method for Anti-inflammatory Activity

Autacoids including a large array of physiological substances like histamine, serotonin, bradykinin, substance P, and the group of eicosanoids (prostaglandins, thromboxanes, and leucotrienes), the platelet- activating factor (PAF) as well as cytokines and lymphokines are involved in the process of inflammation and repair. Their discovery makes the use of in vitro studies possible.

5.1.1 ³H-Bradykinin Receptor Binding

Bradykinin produces pain by stimulating A and C fibers in the peripheral nerves, participates in the inflammatory reaction, and lowers blood pressure by

© Springer India 2015

P. Jain et al., *Inflammation: Natural Resources and Its Applications*,
SpringerBriefs in Immunology, DOI 10.1007/978-81-322-2163-0_5

vasodilatation. The ^{3}H-bradykinin receptor binding is used to detect compounds that inhibit binding of ^{3}H-bradykinin in membrane preparations obtained from guinea-pig ileum. Pieces of guinea pig ileum are homogenized with TES buffer. The homogenates are incubated, centrifuged, and washed with buffer. The formed pellets are incubated with various concentrations of ^{3}H-bradykinin. Total binding is determined in the presence of incubation buffer, nonspecific binding is determined in the presence of nonlabeled bradykinin. Membrane bound radio-activities are determined using a liquid scintillation counter. Mainly two types of bradykinin receptor B_1 and B_2 are known. B_1 receptors have been studied in the isolated rabbit carotid artery, canine cultured tracheal smooth muscle cells, and cultured bovine aortic endothelial cells. B_2 receptors have been studied in human fibroblasts, isolated blood vessels from different species, gall bladder, and smooth muscles of guinea pig. B_1 and B_2 receptors play an important role in bradykinin-induced relaxation and contraction of isolated rat duodenum [1, 2].

5.1.2 ^{3}H-Substance P Receptor Binding

Substance P belongs to the trachykinin family of peptides found in CNS. Substance P is widely distributed in the central and peripheral nervous systems. It causes vasodilatation and plasma extravasations peripherally. In the midbrain it facilitates dopaminergic neurotransmission in response to stress whereas excites dorsal neurons in the spinal cord in response to noxious stimuli. In vitro studies include antagonist of substance P for anti-inflammatory and analgesic activity. The porcine brains obtained from slaughter house are homogenized with Tris-HCl buffer. The homogenates are incubated, centrifuged, and washed with buffer. The formed pellets are incubated with various concentrations of ^{3}H-Substance P. Total binding and nonspecific binding are determined in the absence or presence of unlabeled substance P. Bound radio-activities are determined using a liquid scintillation counter [1].

5.1.3   In Vitro Assay for Polymorphonuclear Leukocyte Chemotaxis

Leukocyte attraction towards infected or inflamed site is an important aspect of host defense mechanism. This method is widely employed to measure the chemotactic effects on polymorphonuclear leukocytes. In this method, multiple tissue-culture plates are utilized and sandwiched. The upper plate is positioned over lower plate and fastened with bolts. The upper compartments are filled with 0.3 ml of polymorphonuclear leukocytes (PMN) suspension. The assembly is incubated; the upper compartment is decanted and the lower plate is centrifuged and decanted. The pellet of PMNs is dispersed in phosphate buffer saline containing EDTA. Each well containing PMN suspension is determined with micro-plate reader [1, 3].

5.1.4   In Vitro Assay of Arachidonic Acid Metabolism

Arachidonic acid is released from the cellular phospholipids fraction by the action of phospholipase A_2, and subsequently metabolized via two major routes: the cyclo-oxygenase pathway yielding the primary prostaglandins and thromboxane, and the 5-lipoxygenase pathway yielding the leukotrienes. Thromboxanes, prostaglandins, and leukotrienes play a pathophysiological role in many diseases. Inhibitors of the 5-lipoxygenase pathway have attracted considerable attention as potential anti-inflammatories with high potency. Along with that, inhibition of prostaglandins by inhibiting COX-1 and COX-2 (cyclooxygenases) are also involved in maintaining vital functions in vascular hemostasis, gastric mucosa, and kidney [1].

5.1.5   In Vitro Assay for Induced Release of Cytokines from Human White Blood Cells

Cytokines are highly potent peptides that are involved in numerous cellular processes, such as inflammation, immunological responses, and many others. Cytokines are synthesized endogenously upon stimulation by infection or injury. It is comprised of several inflammatory substances like interleukin-1 (IL-1), IL-6, IL-8, tumor necrosis factor (TNFα), and many others. Blocking IL-1 or TNF has been highly successful in patients with rheumatoid arthritis or inflammatory bowel disease. Different models are based to detect compounds that interact with cytokines release from human mononuclear blood cells [1].

5.1.6   Flow Cytometric Analysis of Intracellular Cytokines

Flow cytometry is a powerful analytical technique in which individual cells can be simultaneously analyzed for several parameters, including size and granularity, as well as the expression of surface and intracellular markers defined by fluorescent antibodies. Fluorescent anti-cytokine and anti-chemokine monoclonal antibodies are very useful for the intracellular staining and multiparameter flow cytometric analysis of individual cytokine-producing cells within mixed populations. Multicolor immunofluorescent staining with antibodies against intracellular cytokines and cell surface markers provides a high resolution method to identify the nature and frequency of cells which express particular cytokines [4, 5].

5.1.7   Binding to Interferon Receptors

The interferons (IFNs) are biological agents interfering with virus replication. They are a family of secreted proteins occurring in vertebrates and can be classified as cytokines. The IFNs are multifunctional and are components of the host defense against viral and parasitic infections and certain tumors. They affect the functioning of the immune system in various ways and also affect cell proliferation and differentiation. Interferon binds to receptors on the cell surface and induces the synthesis of specific proteins. Analysis of this binding data using different programs is the basis of in vitro models [1].

5.1.8 Screening for Interleukin-1 Antagonists

Interleukin-1-α and -β are potent regulators of inflammatory processes. The naturally occurring interleukin-1 receptor antagonist (IL-1ra) is effective in vitro and in vivo in modulating biological responses to IL-1. Using a combination of anion exchange, gel filtration, and reverse-phase HPLC, three species of native IL-1ra can be identify [1, 6].

5.1.9 Inhibition of Interleukin-1β Converting Enzyme (ICE)

Interleukin-1β-converting enzyme (ICE) is also known as caspase-1; it was the protein identified for programmed cell death (apoptosis) [7]. It is cysteine protease that processes immature pro IL-1 into active IL-1β. IL-1β is a proinflammatory cytokine that mediates many of the physiological and behavioural response to inflammation. Inhibition of IL-1β formation is an approach for treatment of inflammatory disorders such as rheumatoid arthritis [8, 9].

5.2 In Vivo Method for Anti-inflammatory Activity

The inflammatory process involves a series of events that can be elicited by numerous stimuli, e.g., infectious agents, ischemia, antigen-antibody interactions, and chemical, thermal, or mechanical injury. The response is accompanied by the clinical signs of erythema, edema, hyperalgesia, and pain. Different pharmacological methods have been developed for testing acute and sub-acute inflammation, proliferative phase of granuloma formation, and testing of immunological factors.

5.2.1 UV Induced Erythema in Guinea Pig

This test was performed to observe the efficacy of phenylbutazone or other nonsteroidal anti-inflammatory agents to delay the development of ultraviolet erythema on albino guinea pig skin. Erythema is a skin condition characterized by redness or rash. It is caused by a reaction to sunlight and tends to occur when infection or a medication increases your sensitivity to ultraviolet radiation. The shaved skin of guinea pigs are exposed to ultraviolet radiation for 2 min and erythema is scored 2 and 4 h after exposure. The test compound is administered 30 min before ultraviolet exposure. Erythema is scored from 0 to 4 according to severity of injury. The test has the advantage of simplicity but needs training of the investigators. This model is useful to measure the vasodilatory phase in the inflammatory reaction rather to study the duration of the anti-inflammatory effect [1].

5.2.2 Carrageenan Induced Paw Edema Model

This is one of the most commonly employed technique to evaluate the anti-inflammatory effect of the test compound based upon the ability of such agents to inhibit the edema produced in the hind paw of the rat after injection of a carrageenan or other phlogistic agent. Phlogistic agents may be formaldehyde, dextran, egg albumin, kaolin, or brewer's yeast. In this technique, the hind paw is measured and compared with test group animals before and after induction of edema. Rats are challenged by a subcutaneous injection of 0.05 ml of 1 % solution of carrageenan into the plantar side of the left hind paw. The increase of paw volume after 3 or 6 h is calculated as percentage compared with the volume measured immediately after injection of the irritant for each animal. This method has been proven to be suitable for screening purposes as well as for more in-depth evaluations. Depending on the irritant steroidal and nonsteroidal anti-inflammatory drugs, serotonin antagonists are active in the paw edema tests [1, 10].

5.2.3 Carrageenan Induced Pleurisy

Pleurisy is a well-known phenomenon of exudative inflammation that can be induced by several irritants, such as histamine, bradykinin, prostaglandins, mast cell degranulators, dextran, enzymes, antigens, microbes, and irritants, like turpentine and carrageenan. Carrageenan-induced pleurisy in rats is considered to be an excellent acute inflammatory model in which fluid extravasation, leukocyte migration, and the various biochemical parameters involved in the inflammatory response can be measured easily in the exudate. 0.1 ml of 2 % carrageenin solution is injected into the pleural cavity through the incision made into the skin under the right arm between the seventh and eighth rib. The wound is closed with a Michel clip. 1 ml of heparinized Hank's solution is injected into the pleural cavity by making a cut in the body wall to gain access in the pleural cavity and the cavity is gently massaged to mix the contents. The fluid is aspirated out of the cavity using a pipette and collected in a graduated plastic tube. This model has been accepted as a reliable method to study acute and sub-acute inflammation allowing the determination of several parameters simultaneously or successively with anti-inflammatory activity of the drug [1, 10].

5.2.4   Glass Rod Granuloma

This method is used to observe chronic proliferative inflammation along with measurement of chemical composition and mechanical properties of the newly formed connective tissue. A glass rod in inserted into the subcutaneous tunnel by an incision in the caudal region of cranium. The rod remains in situ for 20 or 40 days. After that, the animals are sacrificed under CO_2 anesthesia followed by extraction of glass rod. The glass rod is now prepared with surrounding connective tissue and the wet weight of granuloma tissue is recorded. For measurement of the mechanical properties the specimens are fixed into the clamps of the Instron instrument allowing a gauge length of 30 mm. Additionally, biochemical analyses, such as determination of collagen and glycosaminoglycans, can be performed [1].

5.2.5   In Vivo Inhibition of Leukocyte Adhesion to Rat Mesenteric Venules

Reversible adherence of leukocytes to endothelium, basement membranes, and other surfaces is an essential event in the establishment of inflammation. Leukocyte adhesion to vessel walls greatly increased during inflammation and collision behaves elastically. Adhesion initiates rolling of leukocytes, but further increase in adhesion slows down the rolling process and cells coming to a complete stop and their migration out of the vessel. This process can be observed by preparing a mesenteric venule of an anesthetized rat and following the flow and rolling of leukocytes by means of a microscope. In this test procedure, adhesion of leukocytes, to the vessel wall, is artificially induced by the application of the formyl-methionyl peptide fMet-Leu-Phe (FMLP). Formyl peptides are released from bacteria and mitochondria of damaged tissue, so these peptides provide a specific signal marking the presence of invading bacteria or tissue damage. The density of FMLP receptors ranges from 10^4 to 10^5 per cell, depending on the cell type. Activation of leukocytes through this receptor results in rapid expression of preformed L-selectin (LECAM-1) on the cell surface which causes the cells to roll along the endothelial surface [1].

5.2.6   Oxazolone-Induced Ear Edema in Mice

This is a model of delayed contact hypersensitivity that permits the quantitative evaluation of topical and systemic inflammation. In the test procedure before each use of fresh 2 % solution of oxazolone, mice are sensitized by application of 0.01 ml of halothane on the inside of both ears. The mice are challenged 8 days later again under anesthesia by applying 0.01 ml 2 % oxazolone solution to the inside of the right ear, and the left ear remains untreated. After that animals are sacrificed under anesthesia and a disc of 8 mm diameter is punched from both sides. The discs are immediately weighed on a balance. The weight difference is an indicator of the inflammatory edema [1, 10].

5.2.7   Croton-Oil Ear Edema in Rats and Mice

The method is useful for evaluation of anti-inflammatory topical steroids, especially in the modification when thymus weight is determined simultaneously. In this method, standard and test compounds are dissolved in irritant solution made up of croton oil with concentration of 0.03–1 mg/ml for mice and slightly greater for rats. Four hours after application of compounds ears are removed and the weight difference of punched ears is determined [1, 10].

References

1. Vogel HG (2008) Drug discovery and evaluation: pharmacological assays, vol-I, 3rd edn. Springer, Berlin/Heidelberg/New York
2. Rhaleb NE, Carretero OA (1994) The role of B1 and B2 receptors and of nitric oxide in bradykinin-induced relaxation and contraction of isolated rat duodenum. Life Sci 55:1351–1363
3. Watanabe K, Kinoshita S, Nakagawa H (1989) Very rapid assay of polymorphonuclear leukocyte chemotaxis in vitro. J Pharmacol Methods 22:13–18
4. Jung T, Schauer U, Heusser C, Neumann C, Rieger C (1993) Detection of intracellular cytokines by flow cytometry. J Immunol Methods 159:197–207
5. Sander B, Andersson J, Andersson U (1991) Assessment of cytokines by immunofluorescence and the paraformaldehyde saponin procedure. Immunol Rev 119:65–93
6. Schreuder HA, Rondeau JM, Tardif C, Soffientini A, Sarubbi E, Akeson A, Bowlin TL, Yanofsky S, Barrett RW (1995) Refined crystal structure of the interleukin-1 receptor antagonist. Presence of a disulfide link and a cis-proline. Eur J Biochem 227:838–847
7. Yuan JS, Shaham S, Ledoux S, Ellis HM, Horvitz HR (1993) The C. elegans cell death gene ced-3 encodes a protein similar to mammalian interleukin 1β-converting enzyme. Cell 75:641
8. Dinarello CA (1996) Biological basis for interleukin-1 in disease. Blood 87:2095–2147
9. Tatsuda T, Cheng J, Mountz JD (1996) Intracellular IL-1β is an inhibitor of Fas-mediated apoptosis. J Immunol 157:3949–3957
10. Parmar NS, Prakash S (2011) Screening methods in pharmacology. Narosa Publishing House, New Delhi, p 211

Inflammation and Lifestyle

6

Abstract

Inflammation is the cause of most serious illness including heart disease and stroke, many cancers, and arthritis. Many lifestyle factors, including diet, stress and exposure to environmental and household toxins, contribute to elevated levels of inflammation throughout the body that never subside, creating a cellular environment that is favorable to disease propagation. Our diet is one of the leading sources of these chronic illnesses, and changing the diet is the key to prevention and cure. A number of dietary factors, including fiber-rich foods, whole grains, fruits (especially berries), omega-3 fatty acids, antioxidant vitamins (e.g., C and E), and certain trace minerals (e.g., zinc), have been documented to reduce blood concentrations of inflammatory markers. The best way to correct and eliminate inflammation is to improve comprehensive lifestyle and dietary changes rather than taking pharmaceutical drugs, the latter of which can cause unintended harm in the form of damaging side effects. Intensive lifestyle interventions involving both exercise and diet appear to be most effective. This chapter includes recent news and advances based on anti-inflammatory diets, exercise, and a healthy lifestyle. Several health benefits of natural sources like vegetables, fruits, fishes, beans, and oil are also summarized in this chapter.

6.1 Inflammation News

Inflammation is the first response of your immune system to an infecting organism or foreign material that trips a switch leading to a cascade of biochemical events. Fluid floods the infected area along with a dozen of different chemicals including white blood cells and histamine all capable of producing the signs of inflammation redness, swelling, and pain. When infection occurs, they trigger widespread inflammatory reactions by either directly attacking tissues or creating allergic reactions along with the production of histamine. Various household things and means are available in our surroundings which are easily accessible and implementable to fight

© Springer India 2015
P. Jain et al., *Inflammation: Natural Resources and Its Applications*,
SpringerBriefs in Immunology, DOI 10.1007/978-81-322-2163-0_6

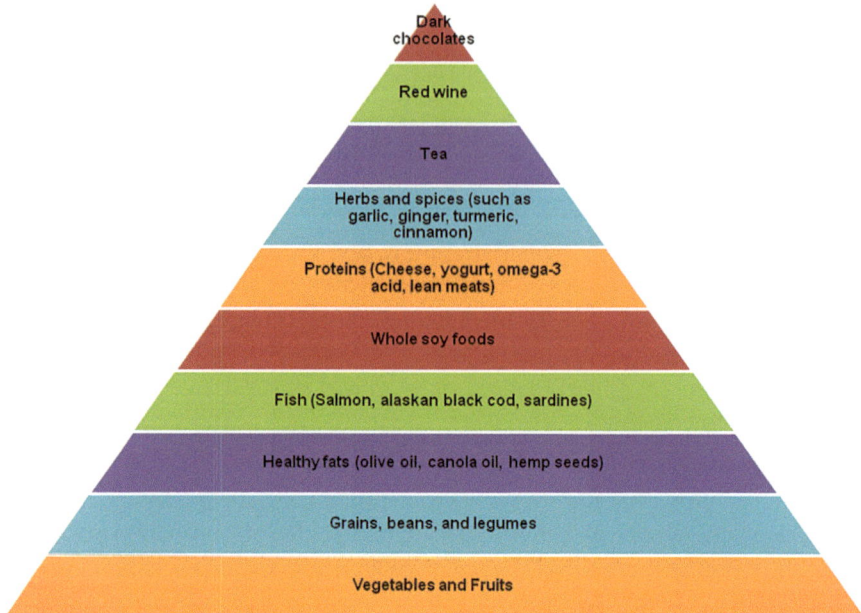

Fig. 6.1 Anti-inflammatory edible's pyramid

against inflammation. Several natural food materials, oils, drinks, and exercises are suggested by researchers and nutritionists useful in day-to-day life as anti-inflammatory sources (Fig. 6.1). Hippocrates also said "let your food be your medicine."

Researchers suggest exercise is more important than diet when working to decrease inflammation. In particular, aerobic exercise appears to improve the balance between pro- and anti-inflammatory markers. In addition to exercise and diet, there are supplements that can help to reduce inflammation including a multivitamin, and fish oil helps to decrease inflammation and the production of free radicals which are damaging to the body. Probiotics are also a good choice because they act as an anti-inflammatory in the gut by introducing good, necessary bacteria.

Making these changes in your life can be overwhelming in reduction of inflammation. Introducing one change at a time will increase the likelihood that you will maintain your new healthy lifestyle. Americans take food high in calories and fat which contain large percentage of sugar, refined flour, and trans fats and act as pro-inflammatory inducers. Refined grains, candy, soda, sugary cereals, processed meats, and high fat meats can be considered foods that promote inflammation. An anti-inflammatory diet is one that is low in processed foods and high in fresh fruits and vegetables, seeds, sprouts, nuts, and super foods. Maca, spirulina, purple corn, wheatgrass, coconut butter, and raw chocolate are a few of the health-promoting super foods that are gaining international interest.

6.1.1   Fish Oil

There are a number of scientifically proven reasons why supplementing with high-quality fish oil is an important part of a healthy diet. Brain, joints, muscles, heart, and skin all rely on the omega-3 fatty acids found in fish oil, without these, body is much more prone to diseases. Fish oil contains two powerful omega-3s, eicosapentaenoic acid (EPA) and docosahexaenoic acid (DHA), that have been shown in numerous scientific studies to quell inflammation. When taken regularly, fish oil that contains a high ratio of both EPA and DHA can help lower the inflammatory response associated with conditions like rheumatoid arthritis, lupus, and chronic back pain, as well as protect against the development of these and other diseases that often result from chronic inflammation [1]. The cardioprotective benefits of fish oil are perhaps its most well-known attributes, as omega-3 fatty acids have been shown time and time again to protect both the heart and cardiovascular system against debilitating illness. Clinical evidence shows that omega-3s protect against high cholesterol by reducing systemic inflammation, as well as balance blood pressure and protect against heart disease. Supplementing with fish oil, in other words, can remedy a host of cognitive abnormalities like chronic "brain fog," depression, neurological disorders, and dementia. Metabolically, omega-3s protect nerves, cells, and various other components of the brain against stress and inflammation, which in turn helps protect memory, hormone production, and nervous system function.

6.1.2   Green Tea

Green tea may help people with rheumatoid arthritis avoid inflammation and joint damage. The researchers cultured synovial fibroblasts from rheumatoid arthritis patients and then exposed the cells to epigallocatechin-3-gallate (EGCG), a naturally occurring compound in green tea. They found that EGCG blocked two potent molecules that cause the bone breakdown in rheumatoid arthritis-affected joints. EGCG blocks production prostaglandin E2, which causes joint inflammation [2].

6.1.3   Plumeria Acuminate

The herb plumeria acuminate commonly known as perungalli has been widely used in ayurvedic medicine, and it may be a potent treatment for both acute and chronic inflammation. Perungalli has been traditionally used in ayurvedic medicine in Southern India and Sri Lanka to treat a wide array of health disorders. The plant material is often used to treat diarrhea and itch, while the milky juice of the plant is used to treat rheumatism and inflammation. The bark is traditionally used as a plaster to cover inflammation or hard tumors, and the leaves reportedly have purgative and anti-inflammatory effects [3].

6.1.4 Pyridoxine

Vitamin B6 or pyridoxine is one of the eight B vitamins of B complex. Each one has an independent but coexisting function. Vitamin B6 is involved with metabolizing amino acids and the synthesis of neurotransmitters serotonin and norepinephrine in addition to the sleep hormone melatonin. Vitamin B6 also helps vitamin B12 to produce red blood cells. It also manages to convert noncarbohydrate sources, such as proteins and lipids, into glucose for cellular metabolic energy. Foods high in vitamin B6 and other B vitamins include brewer's yeast, bee pollen, bell peppers, mushrooms, turnip greens, summer squash, tuna, cod, turkey or chicken, and all the other cruciferous vegetables, such as kale, broccoli, and cauliflower [4].

6.1.5 Meditation

Mindfulness meditation techniques may help to reduce the symptoms of chronic inflammatory conditions such as rheumatoid arthritis, asthma, and inflammatory bowel disease. Mindfulness meditation-based techniques have long been used for stress management and to relieve the symptoms of chronic pain. These techniques typically consist of focusing the attention on the breath, on sensations in the body, and on thoughts or feelings while sitting, walking, or taking part in a body practice such as yoga [5].

6.1.6 DMSO

DMSO (dimethyl sulfoxide) is a natural product from trees that relieves muscle and joint pain, treats bladder inflammation, and protects healthy cells from cancer and chemotherapy drugs. DMSO was on track to becoming a major cancer treatment and promising cure for other ailments. It is approved as an anti-inflammatory treatment for horses. It is also used as a prescription medication for bladder inflammation in people. It can also stop or slow the development of cancers, such as breast, skin, bladder, colon, and ovarian cancer. Some people use it for cancer prevention. DMSO is used to help patients in withdraw from conventional cancer treatment and is promoted as an immune system booster.

6.1.7 Niacin-Bound Chromium (Cr-P)

Inflammation is a normal immune reaction, but chronic inflammation has been linked to a variety of severe health problems, including cardiovascular disease. Because people with diabetes are already predisposed to cardiovascular disease, inflammation is a symptom of particular concern in diabetic patients. Niacin-bound chromium may prove to be effective in preventing chronic inflammation [6].

6.1.8 Olive Oil

Polyphenols from extra virgin olive oil have been shown to significantly reduce the expression of genes that trigger systemic inflammation and can be used along with natural diet to lower the risks from cardiovascular disease. Leafy greens and raw green vegetables are packed with folate, which is known to lower levels of circulating homocysteine that increases risk of coronary artery disease and heart attack. Olive oil contains powerful antioxidants that neutralize free radicals before they can deteriorate normal metabolic function. Extra virgin olive oil is packed with antioxidant compounds and squalene that directly regulate genes that trigger inflammation in the body. The oil was found to reverse the deleterious effect of inflammation caused by stress, obesity, high blood pressure, and blood glucose. Extra virgin olive oil turns off multiple inflammatory genes that are activated as a consequence of metabolic syndrome, effectively providing a protective shield against cardiovascular disease and other chronic illnesses driven by persistent inflammation. Systemic inflammation represents a serious health concern to an aging population and those at increased risk for cardiovascular disease. A solid body of science confirms the health-promoting effects bestowed by a raw diet of leafy green vegetables and powerful antioxidants found in extra virgin olive oil [7].

6.1.9 Zinc

Zinc helps control infections by gently tapping the brakes on the immune response in a way that prevents out of control inflammation that can be damaging and even deadly. Scientists have demonstrated how a specific protein ushers zinc into key cells that stimulate a critical immune response to fight against infection. The mineral interacts with a cellular process that neutralizes infection and helps balance the normal immune response. Researchers found that when a pathogen is detected, a series of complex responses occur to wake the innate immune response utilizing the nuclear-factor kappa beta pathway (NF-kB). The team showed that once NF-kB is activated, a gene is expressed that allows zinc to be ushered from the bloodstream into the cell where it can bind with proteins that block the activity of the pathogen and halt excess inflammation [8].

6.1.10 Pomegranate

Pomegranate extract may inhibit the chronic inflammation linked with a variety of health problems such as heart disease and arthritis. Pomegranate is known to be high in antioxidants, including punicalagins and punicalins, and antioxidants are known to help reduce inflammation in the body. The researchers found that the activity of both inflammation markers (COX-1 and COX-2) in the rabbits given pomegranate extract was significantly reduced [9].

6.1.11 Yoga Reduces Inflammation

Yoga is no longer just a gentle stretching for aging hippies. Ongoing research shows that this ancient ritual reduces inflammation. Yoga reduces the amount of cytokine interleukin-6 (IL-6) in the blood.

6.1.12 Fermented Food and Beverages

Eating more foods and drinking more beverages that are rich in probiotic bacteria, that is, bacteria that promote a healthy, disease-fighting ecosystem inside your digestive tract, is one of the most effective ways to naturally fight inflammation. Since probiotics are vital for the body to effectively break down foods and make them more bioavailable and digestible, they can also help ease the digestive burden brought about by the modern food supply, which is largely responsible for creating inflammation inside the body. Popular probiotic beverages include kombucha tea, unpasteurized apple cider vinegar (ACV), and water kefir.

6.1.13 Omega-3 Fatty Acids

These are natural lubrication for the systems of your body; omega-3 fatty acids like the kinds found in wild fatty fish, hemp and chia seeds, walnuts, and pastured eggs and meats are powerful inflammation fighters. Studies have shown that omega-3s reduce oxidative stress throughout the body and minimize inflammation in the brain, cardiovascular system, and elsewhere, reducing your risk of developing other serious diseases. High-quality fish oils like Green Pasture and Carlson, as well as hemp oil, chia oil, spirulina, pumpkin seed oil, and walnut oil, are all excellent sources of omega-3s. These foods are also all-around nutritional powerhouses that help promote an internal habitat that is unfavorable for inflammation and disease. Low-fat diets are another cause of both inflammation and chronic disease, as the body needs regular intake of healthy fats to keep the circulatory system in good health and maintain healthy blood flow. Consuming more healthy saturated fats in the form of coconut oil, pastured meats and butter, and natural lard cannot only help ease inflammation but also strengthen your bones, improve lung and brain function, and modulate nervous system function.

6.1.14 Nopal Cactus Fruit

Uniquely rich in powerful bioflavonoid nutrients known as betalains, nopal cactus fruit is another must-have anti-inflammatory food that is both delicious and easy to incorporate into your diet. Members of the quercetin family, betalains have been shown to help neutralize the free radicals responsible for triggering inflammation, as well as provide lasting protection against oxidative damage [10].

6.1.15 Maqui Berry

Maqui berry is a vibrant purple fruit that grows abundantly across fields and hillsides in southern Chile. Maqui contains 300 % more anthocyanins and 150 % more polyphenols than any known food or drink, including wine. Maqui is a potent anti-inflammatory due to presence of anthocyanin called delphinidin. Since chronic inflammation contributes to a host of degenerative diseases like arthritis, diabetes, heart disease, and cancer; consuming ample quantities of inflammation curbing maqui is a smart choice. Rich in antioxidants, maqui counteracts free radicals and radiation that contribute to aging. Several other diseases like tumors, hemorrhoids, diabetes, colon cancer, fevers, and diarrhea also respond well to maqui berry [11, 12].

6.1.16 Astaxanthin

Astaxanthin is a red pigment found in different strains of algae, phytoplankton, and plants. Because these organisms constitute the base of many food chains, the pigments can also be found in some animals as well, i.e., salmon and trout. Astaxanthin, a substance known to be the most potent antioxidant, has a natural and scientifically proven anti-inflammatory ability. The fact that astaxanthin is natural and has no side effects makes it very attractive to people as a health supplement. The red pigment is also beneficial for eyes, central nervous system, brain, skin, and immune system and for increasing sports performance and recovery [13].

6.1.17 Bad Eating Habits

Bad eating habits or overeating is a self-perpetuating habit which has devastating health effects. The habit sets off a series of events which short circuit the normal signals the body gives to regulate eating habits. All this leads to wide spread inflammation, contributing to chronic diseases like arthritis, diabetes, heart disease, and cancer. A protein called IKKbeta/NK-kappaB is normally present in the hypothalamus in large quantities. However, it is not generally released into body tissues. This substance is used by macrophages and leukocytes, triggering inflammation throughout the body. Your choice of foods, beyond fat content, can affect the inflammation level in your body. Regular consumption of anti-inflammatory foods may lead to reduction of inflammation on your body. All processed and junk foods contain inflammatory food-like substances. Trans-fatty oils, high fructose corn syrup (HFCS), large doses of sugar, artificial sweeteners, processed salt, and processed white flours are all inflammatory [14].

6.1.18 Celery

A nutrient found in celery has been shown highly effective against inflammation and cancer. Researchers determined that luteolin inhibits lipopolysaccharide-induced interleukin-6 production in the brain through inhibition of the JNK pathway in the inflammatory response of microglia, brain cells in the central nervous system that are key to the body's immune defense. A recent study indicated that only luteolin and quercetin inhibited the platelet-activating factor and suppressed inflammatory response induced by allergens. Luteolin inhibited the excess production of TNF-alpha, a direct cause of inflammation. Celery also contains a good amount of another highly active bioflavonoid, apigenin, a powerful COX-2 inhibitor able to halt inflammation as effectively as anti-inflammatory drugs. It also exhibits antioxidant and antitumor properties. Celery is an excellent source of immune boosting vitamin C, allowing it to be a fighter of the common cold. Its anti-inflammatory properties have also been found effective against asthma [15].

6.1.19 Chokeberry

Chokeberry is a traditionally used fruit found in eastern deciduous forests of the USA. New research shows chokeberries have unusually high levels of anthocyanins that are powerful antioxidants and appear to have potent anti-inflammatory properties that improve blood sugar and the function of insulin and also halt excessive weight gain. Chokeberry extract lowered expression of the gene coding for interleukin-6 (IL-6), a protein that normally triggers inflammation following trauma or infection and that is thought to play a role in the development of a host of human diseases including cancer, diabetes, arthritis, and atherosclerosis [16].

6.1.20 Fats and Cholesterol

Body requires both saturated fat and cholesterol for proper metabolism, brain health, hormone balance, and cellular homeostasis. Without these two important nutritional components, a cascade of health problems can ensue, including debilitating brain conditions like Alzheimer's and Parkinson's [17].

6.1.21 Dark Chocolate and Pistachios

Increase in fats and sugars in chocolate processed from cocoa for semisweet or dark chocolate countered the cacao benefits. C-reactive protein (CRP) is a protein marker produced by the liver that indicates inflammation is occurring throughout the body as the CRP count increases. The CRP blood test is used as an early warning for cardiovascular disease (CVD). Nuts, especially almonds and walnuts, contain beneficial fatty acids for heart health. dark chocolate and pistachios may reduce CRP level [18].

6.1.22 Cilantro

Cilantro is a medicinal herb whose seeds are known as coriander. Seeds contain anti-inflammatory and detoxification properties. Cilantro contains two specific compounds known as cineole and linoleic acid that both possess antiarthritic and antirheumatic properties. Cilantro also contains a substance known as dodecenal that is twice as powerful as the antibiotic drug gentamicin at fighting infection and eradicating harmful microbes from the body. Cilantro is also a natural antiseptic that can help wounds heal more quickly and is a natural chelator of heavy metals from the body. Cilantro might be the natural solution if you suffer from a chronic inflammatory disease, which can manifest itself as arthritis, heart disease, brain fog, fatigue, and irritable bowel syndrome [19].

6.1.23 Resveratrol and Quercetin

Resveratrol, a phytochemical found in red grapes, grape juice, and red wine, has been shown to prolong life in yeast and animals because of its anti-inflammatory and antioxidant properties. Resveratrol and quercetin have been shown to alter levels of inflammation by interacting with DNA sequences at the genetic level. The two super nutrients influence production of the inflammatory marker TNF-α and can help to prevent systemic inflammation and insulin resistance. Resveratrol is known to interfere with the NF-kappaB signaling pathway and to exert a powerful anti-inflammatory signal that can calm the slow-burning flames that contribute to most chronic diseases. Diabetes damage is caused when free radicals attack pancreatic beta cells leading to reduced levels of insulin secretion. The researchers determined that quercetin influences the process that causes instability to the coronary artery walls by attenuating the inflammatory process [20].

References

1. Benson J (2012) Five ways fish oil protects your heart and body. Natural News
2. Gutierrez D (2007) Green tea found to ease inflammation, arthritis pain. Natural news. http://www.naturalnews.com/022435_green_tea.html#ixzz35xF0tCYl
3. Fraser J (2006) Little-known perungalli herb from India shows powerful anti-inflammatory effects in latest clinical trials. Natural news
4. Louis PF (2013) Low vitamin B increases inflammation and causes oxidative stress: eat these foods to boost your levels. Natural news
5. Gutierrez D (2013) Mindfulness meditation shown helpful against chronic inflammation, pain. Natural news
6. Gutierrez D (2008) New study shows Niacin-Bound Chromium Benefits Diabetics. Natural news
7. Phillip J (2011) Olive oil and raw green diet lower systemic inflammation and improve cardiovascular health. Natural news
8. Phillip J (2013) Zinc helps fight infection and inflammation by boosting the body's immune response. Natural news

9. Gutierrez D (2008) Pomegranate extracts found to inhibit inflammation. Natural news
10. Benson J (2013) Reduce widespread inflammation in your body with these foods. Natural news
11. Kilham C (2011) Maqui berry: the Newest Superfruit. Fox News
12. Wright C (2012) Maqui – a rebellious little berry that tames inflammation, prevents cancer and supports a healthy heart. Natural news
13. Gabriele D (2011) Astaxanthin is a top natural anti-inflammatory. Natural news
14. Marshall M (2008) Bad eating habits produce inflammation and changes in behavior. Natural news
15. Minton BL (2008) Celery works great for inflammation, gout, cancer, and high blood pressure. Natural news
16. Baker S (2010) Chokeberry extract found to stop weight gain, regulate blood glucose and halt inflammation. Natural news
17. Benson J (2013) Cholesterol is not the enemy: it's inflammation that's making you fat and killing you slowly. Natural news
18. Louis PF (2009) Dark chocolate and pistachios: the power snack duo that fights inflammation and heart disease. Natural News
19. Benson J (2012) Detoxify heavy metals and soothes chronic inflammation with cilantro, a powerful superherb. Natural news
20. Baker SL (2010) Resveratrol promotes health and longevity; study shows it suppresses inflammation, free radicals. Natural news

Summary and Discussion

Inflammation is a localized, defensive response of the body to injury, usually characterized by pain, redness, heat, swelling, and, depending on the extent of trauma, loss of function. Inflammatory response serves as the body's protective mechanism to treat infection, remove bacteria and toxins, and destroy foreign material at the site of trauma or injury. Due to this response, the spreading of invading substances is halted and the injured area prepares for regeneration or repair [1]. The inflammatory response protects the body against a variety of invading pathogens and foreign matter and should not be confused with an immune response, which reacts to specific invading agents.

If our body cannot solve the problem with inflammation, then our immune system takes over. Inflammation can occur within seconds in response to some kind of injury or attack on the body. Specifically, the inflammation process is an extremely complex and highly evolved system. Effects like redness, swelling, heat, and pain occur due to release of inflammatory mediators that are released after a detected injury or attack. Some of the known mediators include prostaglandins, tumor necrosis factors, interleukins, and nitric oxide. All of these different substances play a role in a complex, coordinated, and balanced defense of the body. The arterioles and capillaries immediately dilate, thus allowing more blood to flow to the injured area.

After dilation, the blood vessel becomes more permeable, allowing plasma and circulating defensive substances such as antibodies, phagocytes, and fibrinogen to pass through the vessel wall to the site of the injury. The blood flow to the area decreases and the circulating phagocytes attach to and digest the invading pathogens. Healing takes place, unless the body's defense system is compromised by a preexisting disease or a weakened condition. Inflammation is described as acute or chronic, depending on how long it lasts.

Acute inflammation is a normal response to physical injury, trauma, or infectious disease. It is an indication that the immune system is functioning accordingly. Chronic inflammation takes long time and is often imperceptible. The problem link with chronic inflammation is the chemical signaling, which makes the immune system response continue unnecessarily attacking healthy tissue, creating a vicious cycle of more inflammation and cellular destruction. Even if the immune response is perceptible, frequent occurrences can act as chronic inflammation and destroy healthy tissue intermittently. All this inflammation leads to heart disease, diabetes

© Springer India 2015
P. Jain et al., *Inflammation: Natural Resources and Its Applications*,
SpringerBriefs in Immunology, DOI 10.1007/978-81-322-2163-0

II, and a host of neurodegenerative diseases that have become more common over the last 50 plus years [2]. It makes sense to limit or break up those patterns of chronic inflammation, which can lead to disease and aging more quickly. A primary cause of chronic illness for many people is systemic inflammation and particularly the kind that goes unaddressed or even mostly unnoticed until it eventually progresses into a full-blown chronic disease. Infection from viral and bacterial agents can pose a significant threat to human health, as it fuels the flames of systemic inflammation known to contribute to high blood pressure, bacterial and viral infections, arthritis, acid reflux, premature aging, cancer, and many other common health-related problems [3]. The best way to correct and eliminate inflammation is to improve comprehensive lifestyle and dietary changes rather than taking pharmaceutical drugs, the latter of which can cause unintended harm in the form of damaging side effects. Many lifestyle factors, including diet, stress, and exposure to environmental and household toxins, contribute to elevated levels of inflammation throughout the body that never subside, creating a cellular environment that is favorable to disease propagation [4]. Nature has a number of sources which provide natural compounds and nutrients that help combat inflammation and lower risk of infection and chronic disease. It is well established that obesity is characterized by low-level chronic inflammation [5]. Several theories as to how this inflammation occurs have been proposed, but natural sources including plant, animal, and marine source seem to be the most important factor for treating inflammation. It is increasingly recognized that regular consumption of certain foods leads to diet-induced inflammation, which in turn sets the stage for insulin resistance, leptin resistance, and other conditions that go hand in hand with overweight and obesity [6].

Ridding yourself of body fat may help remove chronic inflammation. It has been discovered that body fat often acts as an endocrine organ, secreting proinflammatory chemicals. Fatty liver is one example, and visceral fats are inflammatory conditions. But eliminating all fat from diet is the wrong approach. Our brains, nerve tissues, and cell walls are composed of fat, even cholesterol. The sunshine to vitamin D3 conversion factor starts with a type of cholesterol in our skin. Based on the knowledge about inflammation and obesity, it is likely that the limited consumption of cellular carbohydrates and proinflammatory antinutrients and proteins is very important in reducing inflammation and promoting weight loss on these diets.

Saturated fats like those found in butter and fat from grass-fed animals, as well as in coconut and palm oils, are actually beneficial for your health, while carbohydrates and oils rich in omega-6 fatty acids are heavy promoters of disease-causing inflammation [7]. One should skip the low-fat diet and start eating more healthy fats in combination with mineral and enzyme-rich whole foods. Omega-3 fatty acids are anti-inflammatory, while omega-6 acids can actually cause inflammation. That is why we need to balance the use of omega-6 with omega-3. If our diet has been predominantly rich in the omega-6 oils, body will have to do more than achieve a balance by increasing omega-3 and drastically reducing the omega-6.

Richest sources of omega-3 fatty acid include oily fish such as herring, sardines, tuna, mackerel, and salmon. Plant sources such as hemp, flax, pumpkin seeds, walnuts and their oils, and high-fiber non-starchy vegetables such as dark leafy salad greens, spinach, kale, tomatoes, broccoli, cauliflower, collard greens, and onions are all that contribute to fight against inflammation [8]. It is recommended by the researchers and nutritionists to avoid consuming inflammation-causing processed foods and pharmaceutical drugs. Dietary polyphenols, found in many edible plants, are being found to have anti-inflammatory properties. Studies on both animals and humans have shown that consumption of polyphenol-rich foods lowers incidences of inflammatory disease [8].

Fermented foods like yogurt, sauerkraut, apple cider vinegar, and fermented vegetables are all excellent examples of probiotic-rich superfoods. These richly fermented foods help to populate gut with beneficial bacteria and ensure healthy and well functioning of digestive tract. Fermented foods and beverages also help prevent harmful pathogens from taking hold within the body. Vitamin D is a powerful natural anti-inflammatory source that is easily accessible through natural sunlight exposure or supplementation, and it is one of the most powerful interventions for deterring inflammation [7].

Fruits rich in phytochemicals like berries, sour cherries, pomegranates, and cranberries all help to remove inflammation. Turnip, a Chinese medicinal herb, and radish both aid digestion by cooling and soothing inflammation and phlegm. Quercetin is a flavonoid and is very powerful. It is found in red grapes, red and yellow onions, garlic, broccoli, and apples. Antioxidant properties in some foods also help fight inflammation by protecting the body from free radicals. Vitamin C-rich foods fall into this category, including carrots, orange winter squash, bell peppers, and tomatoes. Papaya and pineapple also help to reduce swelling and inflammation quickly. Pineapple contains the enzyme bromelain and papaya contains papain. Pain and swelling should go down in 2–6 days. It is recommended by the nutritionists to increase the intake of this food group during inflammation [8]. Animal sources, i.e., lean meats and fishes, are also very important in fighting inflammation. Indian foods involve lots of spices, condiments, oils, vegetables, fruits, and other accessories that contain good anti-inflammatory properties. Turmeric, and its yellowing substance curcumin, is most commonly found in Indian foods like curry and in mustards. Ginger is another flavoring that has many healthful properties, one of which is an anti-inflammatory [8].

Garlic has been the object of much research and has been shown to inhibit the growth of 23 organisms, including bacteria, mold, and yeast. It is also very helpful as an anti-inflammatory agent. But one caution with the use of herbs and spices is that it should be used with medical supervision because their medicinal properties can interfere with the drugs.

Even conventional medicine is now recognizing how pivotal inflammation can be in determining the course of human health. The more conventional medicine learns about inflammation, the more it realizes that inflammation is the precursor to dozens of other serious diseases. By halting the inflammation and eliminating its root cause, one can halt the progression of degenerative disease and begin healing from the inside out.

Many of the natural sources are highly endangered and urgently need to be maintained in their native habitats. Unless we preserve genetic material for propagation from these species now, many will extinct before we can protect and restore habitats for their long-term recovery. The viable plant material, living plant collections, and long-term seed storage can be preserved in order to maximize their potential for future use in our restoration efforts [9].

References

1. Blakemore C, Jennett S (2001) "Inflammation." The Oxford Companion to the Body. Encyclopedia. http://www.encyclopedia.com
2. Louis PF (2012) Banish chronic inflammation to rejuvenate your health – top tips for detoxification and diet overhaul. Natural News. http://www.naturalnews.com/037203
3. Benson J (2013) Reduce widespread inflammation in your body with these foods. Natural News. http://www.naturalnews.com/038915
4. Hunter E (2012) Inflammation is a major reason why you can't lose weight. Natural News. http://www.naturalnews.com/036701
5. Lumeng CN, Saltiel AR (2011) Inflammatory links between obesity and metabolic disease. J Clin Invest 121(6):2111–2117
6. Bastard JP, Maachi M, Lagathu C (2006) Recent advances in the relationship between obesity, inflammation, and insulin resistance. Eur Cytokine Netw 17(1):4–12
7. Benson J (2012) Inflammation is the cause of nearly all disease – here's how to prevent it. Natural News. http://www.naturalnews.com/036722
8. Sherman C (2008) Inflammation: what to eat to reduce your risk of many diseases. http://www.naturalnews.com/022701
9. Kumar MA, Parveen B, Sanjiv K (2009) Plants-herbal wealth as a potential source of ayurvedic drugs. Asian J Tradit Med 4(4)